Lecture Notes in Mathematics

Edited by A. Dold and B. Ec

545

Noncommutative Ring Theory

Papers presented at the International Conference
Held at Kent State University
April 4–5, 1975

Edited by J. H. Cozzens and F. L. Sandomierski

Springer-Verlag
Berlin · Heidelberg · New York 1976

Editors
John Henry Cozzens
Department of Mathematics
Rider College
Lawrenceville, New Jersey 08648/USA

Francis Louis Sandomierski
Department of Mathematics
Kent State University
Kent, Ohio 44242/USA

Library of Congress Cataloging in Publication Data

International Conference on Noncommutative Ring Theory,
Kent State University, 1975.
Noncommutative ring theory.

(Lecture notes in mathematics ; 545)
Bibliography: p.
Includes index.
1. Noncommutative rings--Congresses. I. Cozzens,
John, 1942- II. Sandomierski, Francis L., 1937-
III. Title. IV. Series: Lecture notes in mathematics
(Berlin) ; 545)
QA3.L28 no. 545 [QA251.4] 510'.8s [512'.4]
76-46377

AMS Subject Classifications (1970): 15A69, 15A72, 16A02, 16A08, 16A10, 16A12, 16A18, 16A30, 16A40, 16A46, 16A49, 16A52, 16A64

ISBN 3-540-07985-8 Springer-Verlag Berlin · Heidelberg · New York
ISBN 0-387-07985-8 Springer-Verlag New York · Heidelberg · Berlin

Printed in Germany

Printing and binding: Beltz Offsetdruck, Hemsbach/Bergstr.

TABLE OF CONTENTS

CONTRIBUTORS

John A. Beachy, Northern Illinois University, DeKalb, Illinois

George M. Bergman, University of California, Berkeley, California.

Joe W. Fisher, The University of Texas, Austin, Texas.

K. W. Goodearl, University of Utah, Salt Lake City, Utah

Robert Gordon, Temple University, Philadelphia, Pennsylvania.

Arun V. Jategaonkar, Fordham University, Bronx, New York.

Thomas H. Lenagan, Mathematical Institute, Edinburgh, Scotland.

J. C. Robson, University of Leeds, Leeds, Great Britain.

Mark Teply, University of Florida, Gainesville, Florida.

PREFACE

The volume contains a selection of articles on ring theory presented at the Internation Conference on Noncommutative Ring Theory at Kent State University, April 4-5, 1975.

Thanks are due to Springer-Verlag for publishing these articles and for their generous support.

The conference itself was sponsored by the mathematics department at Kent State University.

We are deeply indebted to Kathy Morrison who accepted the challenge of serving as administrative assistant to us, and in spite of the best efforts of Mother Nature did a remarkable job of managing the conference.

The existence of the proceedings is due to the assistance and cooperation of the authors; the final typescript the efforts of the ubiquitous Ms. Morrison.

Finally we would like to thank Midge Cozzens for her unending patience, kind assistance and hospitality to the participants of the conference.

J. H. Cozzens

F. L. Sandomierski

SOME ASPECTS OF NONCOMMUTATIVE LOCALIZATION

John A. Beachy
Northern Illinois University
DeKalb, Illinois

This paper is expository in nature, although several results (including 1.5, 1.6(2), 2.7 and 3.1) appear to be new, and continues the account of localization for noncommutative rings begun in [4]. In recent papers, Lambek and Michler [15, 16] and Jategaonkar [13] have investigated the localization at a prime or semiprime ideal of a Noetherian ring. By studying the localization at a semiprime Goldie ideal, as in [7], some of their results can be extended to any ring with Krull dimension (see [9] for the definition of Krull dimension). An ideal I of a ring R will be called semiprime (prime) Goldie if the ring R/I is a semiprime (prime) ring with a.c.c. on left annihilators and finite uniform dimension on the left. Any semiprime ideal of a ring with Krull dimension is of this type [9, Corollary 3.4]. The approach in section 3 also makes it possible to extend Small's theorem [19] to give conditions under which the localization at a semiprime Goldie ideal is Artinian.

In Theorem 3.1, the module theoretic characterization of the torsion radical determined by a prime Goldie ideal uses the notion of a strongly prime module, which was introduced in [6]. Section 1 contains a characterization of strongly prime modules, some results on strongly prime rings, and a torsion theoretic characterization of semiprime left Goldie rings. Maximal torsion radicals, used in the latter characterization, are studied in section 2. The Walkers [22] have shown that for a commutative Noetherian ring, maximal torsion radicals correspond to minimal prime ideals, and this result can also be extended [2] to rings with Krull dimension (on either side). In the interests of simplicity, this result, as well as the extension of Small's theorem, is stated for a left Noetherian ring, although the proof is valid in the more general setting.

Throughout the paper, R-Mod will denote the category of unital left R-modules over an associative ring R with identity. A functor σ:R-Mod $\to$ R-Mod is called a torsion preradical if for all modules ${}_RM$, ${}_RN$ and all $f \in \mathrm{Hom}_R(M,N)$, $\sigma M \subseteq M$, $f(\sigma M) \subseteq \sigma N$, (i.e., σ is a subfunctor of the identity) and $\sigma M' = \sigma M \cap M'$ for all submodules $M' \subseteq M$. The submodule σM will be denoted by $\mathrm{rad}_\sigma(M)$. The functor σ is called a torsion radical if in addition $\mathrm{rad}_\sigma(M/\mathrm{rad}_\sigma(M)) = 0$ for all $M \in$ R-Mod. A module ${}_RM$ is called σ-torsion if $\mathrm{rad}_\sigma(M) = M$ and σ-torsionfree if $\mathrm{rad}_\sigma(M) = 0$; a submodule $M' \subseteq M$ is σ-dense if M/M' is σ-torsion and σ-closed if M/M' is σ-torsionfree. The σ-closure of M' in M is defined as the intersection of all σ-closed submodules of M which contain M'.

We will use the notation $\sigma \le \tau$ for torsion preradicals σ, τ such that $\mathrm{rad}_\sigma(M) \subseteq \mathrm{rad}_\tau(M)$ for all $M \in$ R-Mod. A torsion radical is called maximal if it is proper (not the identity functor) and is maximal with respect to the relation $\le$. If ${}_RW$ is an injective module, then for any module ${}_RM$ let $\mathrm{rad}_W(M)$ be the intersection of all kernels of homomorphisms from M into W. Then rad_W defines a torsion radical of R-Mod, and every torsion radical of R-Mod is of this form. The torsion radical $\mathrm{rad}_{E(N)}$ is the largest torsion radical σ for which N is σ-torsionfree, where E(N) denotes the injective envelope of N. Thus for any torsion radical σ, $\mathrm{rad}_\sigma(N) = 0$ iff $\sigma \le \mathrm{rad}_{E(N)}$.

For the torsion radical $\sigma = \mathrm{rad}_W$, where W is an injective module, a left ideal $A \subseteq R$ is σ-closed iff A is the left annihilator of a subset of W. In particular, the ideal $K = \mathrm{rad}_\sigma(R)$ is the left annihilator of W. An ideal K is called a torsion ideal if it is the left annihilator of an injective module, and in this case $K = \mathrm{Ann}(E(R/K))$. This shows that the R-injective envelope of R/K coincides with the R/K-injective envelope of R/K, since in general, if M is an R/I-module for some ideal I, then the R/I-injective envelope of M is $E({}_{R/I}M) = \{x \in E({}_RM) \mid Ix = 0\}$.

For a torsion radical σ, the full subcategory determined by all modules ${}_RM$ such that E(M) and E(M)/M are σ-torsionfree is called the quotient category determined by σ, and will be denoted by R-Mod/σ. The inclusion functor from R-Mod/$\sigma \to$ R-Mod has an exact left adjoint, denoted by Q_σ, and defined by letting $Q_\sigma(M)$ be the σ-closure of $M/\mathrm{rad}_\sigma(M)$ in $E(M/\mathrm{rad}_\sigma(M))$. The module of quotients $Q_\sigma(M)$ will be denoted simply by M_σ.

For any module $M \in$ R-Mod/σ, and any element $m \in M$, the homomorphism $[r \to rm]:R \to M$ defined by multiplication can be extended uniquely to $\rho_m:R_\sigma \to M$. For any element $q \in R_\sigma$, ρ_q can be used to define right multiplication by q, and this induces a ring structure on R_σ. Furthermore, any module $M \in$ R-Mod/σ becomes a left R_σ-module by defining $qm = \rho_m(q)$, for all $q \in R_\sigma$ and $m \in M$. The ring R_σ is called the ring of quotients determined by σ. The quotient category R-Mod/σ is also a quotient category of R_σ-Mod, and the functor Q_σ can be viewed as a functor from R-Mod to R_σ-Mod, although as such it may be only left exact.

The torsion radical σ is called perfect if the following equivalent conditions are satisfied: (1) R-Mod/σ coincides with R_σ-Mod, (2) Q_σ is naturally isomorphic to $R_\sigma \otimes_R -$, (3) $R_\sigma D = R_\sigma$ for every σ-dense left ideal $D \subsetneq R$. The module ${}_RW$ is an injective cogenerator in R-Mod/σ iff W is injective in R-Mod and $\sigma = \mathrm{rad}_W$, and it can be shown that σ is perfect iff W is a cogenerator in R_σ-Mod.

The torsion radical σ is said to be a prime torsion radical if the quotient category R-Mod/σ has an injective cogenerator which is the injective envelope of a simple object in R-Mod/σ, in which case R-Mod/σ has a unique simple object (up to isomorphism). Equivalently, $\sigma = \mathrm{rad}_{E(M)}$ for a monoform module M. (M is monoform if every nonzero homomorphism from a submodule of M into M is a monomorphism, which occurs iff the endomorphism ring of the quasi-injective envelope of M is a division ring [4, Proposition 2.8].) If M is monoform, $\sigma = \mathrm{rad}_{E(M)}$, and $M = M_\sigma$, then σ is perfect iff $R_\sigma/J(R_\sigma)$ is a simple ring and M is the only simple R_σ-module (up to isomorphism). Note that part (a) of [4, Proposition 3.8] should be corrected to include the condition that R_σ has only one isomorphism class of simple modules.

The complete ring of quotients $Q_{max}(R)$ is defined by the torsion radical $\mathrm{rad}_{E(R)}$. For $\sigma = \mathrm{rad}_{E(R)}$, the terms σ-torsionfree, σ-dense, and σ-closed will be abbreviated to torsionfree, dense, and closed, respectively. A torsion radical σ will be called complete if $\sigma = \mathrm{rad}_{E(R/K)}$ for $K = \mathrm{rad}_\sigma(R)$. In this case, since K is a torsion ideal, $E(R/K) = E({}_{R/K}R/K)$ and consequently R_σ can be identified with $Q_{max}(R/K)$.

The singular submodule Z(M) of ${}_RM$ is the set of elements of M whose left annihilator is an essential left ideal of R. This defines a torsion preradical, and the Goldie torsion radical G is the smallest torsion radical which contains it. Then $\mathrm{rad}_{E(R)} \leq G$, with equality iff $Z(R) = 0$, or equivalently, iff $Q_{max}(R)$ is von Neumann regular.

§1. Strongly prime rings and modules.

If ρ is a subfunctor of the identity on R-Mod, then setting $\mathrm{rad}_\tau(M) = M \cap \rho(E(M))$ for all $M \in$ R-Mod defines a torsion preradical τ with $\rho \leq \tau$ and $\tau \leq \sigma$ for all torsion preradicals σ such that $\rho \leq \sigma$. For a module ${}_RN$ the smallest torsion preradical τ such that N is τ-torsion will be denoted by Rad^N, and since any sum of homomorphic images of N must be τ-torsion, it can be shown that $\mathrm{Rad}^N(M) = M \cap \sum_{\alpha\in J} f_\alpha(N)$, where f_α runs through all homomorphisms in $\mathrm{Hom}_R(N, E(M))$. With this definition, it is easy to see that Rad^N is the identity functor iff N generates (in the categorical sense) every injective left R-module. That is, $\mathrm{Rad}^N = 1$ iff N is cofaithful. (A module N is faithful iff it cogenerates all projectives.) The module N is cofaithful iff there exist $x_1, \ldots, x_m \in N$ such that $\mathrm{Ann}(x_1, \ldots, x_m) = 0$, so that R can be embedded in a finite direct sum N^m of copies of N.

A nonzero module ${}_RM$ is called prime if for any left ideal $A \subseteq R$ and any submodule $N \subseteq M$, $AN = 0$ implies $AM = 0$ or $N = 0$. This is equivalent to the condition that $\mathrm{Ann}(N) = \mathrm{Ann}(M)$ for all nonzero submodules $N \subseteq M$.

Definition 1.1. A nonzero module ${}_RM$ is called strongly prime if M is prime and for each submodule $0 \neq N \subseteq M$ and each element $y \in M$ there exist elements $x_1, \ldots, x_n \in N$ such that $\mathrm{Ann}(x_1, \ldots, x_n) \subseteq \mathrm{Ann}(y)$. The ring R is called left strongly prime if the module ${}_RR$ is strongly prime, and an ideal $P \subseteq R$ is called (left) strongly prime if R/P is left strongly prime.

Proposition 1.2. The following conditions are equivalent for a nonzero module ${}_RM$:

(1) M is strongly prime;

(2) For any torsion preradical τ of R-Mod, either $\mathrm{rad}_\tau(M) = 0$ or $\mathrm{rad}_\tau(M) = M$;

(3) M is contained in every nonzero fully invariant submodule of E(M);

(4) For each $y \in M$ and $0 \neq x \in M$ there exist $r_1, \ldots, r_n \in R$ such that $\mathrm{Ann}(r_1x, \ldots, r_nx) \subseteq \mathrm{Ann}(y)$.

Proof. (1) => (2). If $0 \neq \mathrm{rad}_\tau(M)$ for some torsion preradical τ, then for $y \in M$ there exist $x_1, \ldots, x_n \in \mathrm{rad}_\tau(M)$ such that $\mathrm{Ann}(x_1, \ldots, x_n) \subseteq \mathrm{Ann}(y)$. Thus for $x = (x_1, \ldots, x_n) \in (\mathrm{rad}_\tau(M))^n$, $f: Rx \to Ry$ given by $f(ax) = ay$ for all $a \in R$ is well-defined, so $Ry \subseteq \mathrm{rad}_\tau(M)$ since Rx is τ-torsion. Thus $M = \mathrm{rad}_\tau(M)$.

(2) => (3). If $0 \neq N \subseteq E(M)$ is fully invariant, then $\mathrm{Rad}^N(E(M)) = N$, so $\mathrm{Rad}^N(M) = M \cap \mathrm{Rad}^N(E(M)) = M \cap N \neq 0$. Thus we must have $\mathrm{Rad}^N(M) = M$, and so $M \subseteq N$.

(3) => (4). For $0 \neq x \in M$, let N be the sum in E(M) of the homomorphic images of Rx. Then N is fully invariant, so by assumption $M \subseteq N$ and thus $y = \sum_{i=1}^n f_i(r_ix)$ for $r_i \in R$ and $f_i \in \mathrm{Hom}_R(Rx, E(M))$. Therefore $ay = 0$ if $ar_ix = 0$ for all i.

(4) => (1). If $0 \neq N \subseteq M$ and $AN = 0$ for some left ideal $A \subseteq R$, then let $0 \neq x \in N$. By assumption for any $y \in M$ there exist $r_1, \ldots, r_n \in R$ such that $A \subseteq \mathrm{Ann}(r_1x, \ldots, r_nx) \subseteq \mathrm{Ann}(y)$, so $AM = 0$. This shows that M is prime, and the second condition follows immediately. ■

Condition 4 of Proposition 1.2 can be used to show that if ${}_RM$

is strongly prime, then any nonzero submodule of M is strongly prime, and any direct sum of copies of M is strongly prime. Condition 3 then shows that M is strongly prime iff its quasi-injective hull $\overline{M}$ is strongly prime, since $\overline{M}$ is the smallest fully invariant submodule of E(M) which contains M. Furthermore, the proof of (1) => (2) can be used to show that a quasi-injective module is strongly prime iff it is generated (in the categorical sense) by each of its nonzero submodules, and from this it follows that M is strongly prime iff each of its nonzero submodules generates $\overline{M}$.

For the module $_RR$, taking y = 1 in Definition 1.1 shows that R is left strongly prime iff R is prime and every nonzero left ideal is cofaithful. The latter condition implies that R is prime, and is equivalent to the condition that every nonzero ideal of R is cofaithful. The remarks of the preceding paragraph show that R is left strongly prime iff there exists a cofaithful strongly prime left R-module. Furthermore, condition 3 of the next proposition (which follows immediately from Proposition 1.2) shows that R is left strongly prime iff $E(_RR)$ is strongly prime.

Proposition 1.3. The following conditions are equivalent for the ring R:

(1) R is left strongly prime;

(2) $\mathrm{rad}_\tau(R) = 0$ for every proper torsion preradical τ of R-Mod;

(3) E(R) has no nontrivial fully invariant submodules;

(4) If $0 \neq b \in R$, then there exist $r_1, \ldots, r_n \in R$ such that $ar_ib = 0$ for all i implies a = 0.

Strongly prime rings were originally studied by Rubin [17] as absolutely torsion-free rings, using condition 2 of Proposition 1.3. Viola-Prioli [20] showed the equivalence of conditions 1 and 2; Handelman and Lawrence [10] used the name strongly prime for rings which satisfy condition 4, and showed the equivalence of conditions 2 and 4. Proposition 1.2 was used in [6] to show that the endomorphism ring of a finitely generated projective module over a left strongly prime ring is left strongly prime. The following corollary is due to Rubin [17].

Corollary 1.4. If R is left strongly prime, then $Z({}_RR) = 0$ and $Q_{max}(R)$ is a simple, von Neumann regular ring.

Proof. The singular submodule defines a torsion preradical of R-Mod so either $Z(R) = 0$ or $Z(R) = R$, and the latter is impossible since Ann(1) is not essential. Since $Z(R) = 0$, $Q_{max}(R)$ is von Neumann regular and $Q_{max}(R) = E(R)$ as R-modules, which shows that $Q_{max}(R)$ is simple because E(R) has no nontrivial fully invariant R-submodules. ■

Proposition 1.5. Let σ be a torsion radical of R-Mod, with $rad_\sigma(R) = K$ and $rad_\sigma(E(R)) = N$. The following conditions are equivalent:

(1) σ is maximal in the set of proper torsion preradicals of R-Mod;

(2) σ is complete, and for any ideal $A \supsetneq K$ of R, $N \oplus A/K$ is cofaithful;

(3) σ is complete, K is a strongly prime ideal, and there exist $x_1, \ldots, x_m \in N$ such that $K \cap Ann(x_1, \ldots, x_m) = 0$.

Proof. (1) => (3). Since $rad_\sigma(R/K) = 0$, $\sigma \leq rad_{E(R/K)}$, so by assumption we must have equality, and thus σ is complete. If $A \supsetneq K$, then let τ be the smallest torsion preradical such that $\sigma \leq \tau$ and A/K is τ-torsion. Since $rad_\sigma(A/K) \neq A/K$, $\sigma \neq \tau$ and so $\tau = 1$ by assumption. It can be shown that $rad_\tau(E(R)) = rad_\sigma(E(R)) + Rad^{A/K}(E(R))$, so there exist elements $x \in N$, $y_1, \ldots, y_m \in A/K$, and $f_1, \ldots, f_m \in Hom_R(A/K, E(R))$ such that $x + \sum_{i=1}^m f_i(y_i) = 1 \in R$. If $k \in K$ and $kx = 0$, then $k = 0$ since $ky_i = 0$ for all i, so $K \cap Ann(x) = 0$. Furthermore, since $rad_\sigma(E(R/K)) = 0$, $Rad^{A/K}E(R/K) = rad_\tau(E(R/K)) = E(R/K)$ which shows that A/K is cofaithful as an R/K-module. Thus every nonzero ideal of R/K is cofaithful and R/K is left strongly prime.

(3) => (2). By assumption if $A \supsetneq K$, then A/K is cofaithful as an R/K-module, so there exist elements $a_1, \ldots, a_n \in A$ such that $ra_i \in K$ for all i implies $r \in K$. If in addition $rx_i = 0$ for $i = 1, \ldots, m$, then

$r \in K \cap \mathrm{Ann}(x_1, \ldots, x_m) = 0$. This shows that $N \oplus A/K$ is cofaithful.

(2) => (1). If τ is a torsion preradical with $\tau \geq \sigma$ and $\mathrm{rad}_\tau(R/K) = 0$, then $\tau \leq \sigma$ and so $\tau = \sigma$. If $\mathrm{rad}_\tau(R/K) \neq 0$, then let $\mathrm{rad}_\tau(R/K) = A/K$, where A is an ideal of R since $\mathrm{rad}_\tau(R/K)$ is fully invariant in R/K. By assumption $N \oplus A/K$ is cofaithful, and so $\tau \geq \mathrm{Rad}^{N\oplus A/K} = 1$.∎

Theorem 1.6. [3, 6, 10, 18] The following conditions are equivalent for the ring R:

(1) R is semiprime and every faithful left ideal of R is cofaithful;

(2) R is semiprime, and for each ideal A of R there exist elements $a_1, \ldots, a_n \in A$ such that $\ell(A) = \ell(a_1, \ldots, a_n)$;

(3) R-Mod has finitely many maximal torsion radicals (which correspond to the minimal prime ideals of R), and every proper torsion preradical of R-Mod is contained in one of them;

(4) Every proper torsion preradical of R-Mod is contained in a maximal torsion radical;

(5) R is a subdirect product of finitely many left strongly prime rings.

Proof. $\ell(A)$ denotes the left annihilator $\{r \mid rA = 0\}$ of a set A, while $\eta(A)$ denotes the right annihilator.

(1) => (2). If A is an ideal of R, then $\ell(A) \cap A = 0$ since R is semiprime, and $\ell(A) \oplus A$ is faithful. ($B(\ell(A) \oplus A) = 0$ implies $BA = 0$, so $B^2 \subseteq B\ell(A) = 0$ and $B = 0$.) By assumption there exist $x_1, \ldots, x_n \in \ell(A) \oplus A$ such that $\ell(x_1, \ldots, x_n) = 0$, with $x_i = y_i + a_i$, where $y_i \in \ell(A)$ and $a_i \in A$. If $r \in \ell(a_1, \ldots, a_n)$ and $a \in A$, then $arx_i = 0$ for all i since $ry_i \in \ell(A) = \eta(A)$, and so $ar = 0$. Thus $r \in \eta(A) = \ell(A)$, and so $\ell(A) = \ell(a_1, \ldots, a_n)$. Note that $\ell(A) = \eta(A)$ since R is semiprime.

(2) => (3). If $\{A_\alpha\}_{\alpha \in J}$ is any ascending chain of annihilator ideals with $\ell(A_\alpha) \neq 0$ for all α, then by assumption $\ell(\cup_{\alpha \in J} A_\alpha) = \ell(a_1, \ldots, a_n)$

for elements $a_1, \ldots, a_n \in \cup_{\alpha \in J} A$, so $\ell(\cup_{\alpha \in J} A_\alpha) = \ell(A_\alpha) \neq 0$ for some $\alpha \in J$. Applying Zorn's Lemma shows that any proper annihilator ideal is contained in a maximal annihilator ideal, which is then a minimal prime ideal.

If τ is any proper torsion preradical, then $\text{rad}_\tau(R)$ is not cofaithful, so by assumption $\ell(\text{rad}_\tau(R)) \neq 0$. Thus $\text{rad}_\tau(R) \subseteq P$ for a minimal prime ideal P, by the preceding argument, and $\eta(P) \cap \text{rad}_\tau(R) = 0$. Then $\text{rad}_\tau(\eta(P)) = 0$, and so $P = \ell(\eta(P))$ implies that R/P can be embedded in a direct product of copies of $\eta(P)$. This shows that $\text{rad}_\tau(R/P) = 0$ and thus $\tau \leq \text{rad}_{E(R/P)}$.

Let P be a minimal prime ideal of R. Then $\mu = \text{rad}_{E(R/P)}$ is a maximal torsion radical, since if $\mu \leq \sigma$ and $\sigma \neq 1$, then as above, $\sigma \leq \text{rad}_{E(R/P')}$ for a minimal prime ideal P'. Now $\text{rad}_{E(R/P)} \leq \text{rad}_{E(R/P')}$ implies that P' is $\text{rad}_{E(R/P)}$-closed, so P' is the annihilator of a nonzero submodule of $E(R/P)$. It follows that $P' \subseteq P$, so $P' = P$ and $\mu = \sigma$ since P' is a minimal prime. Thus the minimal prime ideals of R are just the maximal annihilator ideals, and these correspond to the maximal torsion radicals of R-Mod.

If $A \neq B$ are minimal annihilator ideals, then $A \cap B = 0$ since $A \cap B$ is an annihilator ideal. Thus $AB = 0$, and from this it follows that the sum in R of all minimal annihilator ideals is a direct sum, say $\oplus_{\alpha \in J} A$. By assumption $\ell(\oplus_{\alpha \in J} A_\alpha) = \ell(a_1, \ldots, a_n)$ for elements $a_1, \ldots, a_n \in \oplus_{\alpha \in J} A_\alpha$, so $\ell(\ {}_{\alpha \in J} A_\alpha) = \ell(\oplus_{i=1}^n A_i)$ for some finite subset $\{A_i\}_{i=1}^n$. If A is a minimal annihilator which is not in the finite subset, then $A(\oplus_{i=1}^n A_i) = 0$, and this implies that $A^2 \subseteq A(\oplus_{\alpha \in J} A_\alpha) = 0$, so $A = 0$.

(3) $\Rightarrow$ (4). Immediate.

(4) $\Rightarrow$ (1). If A is a nonzero ideal of R, then $\text{Rad}^{R/A} \neq 1$, so by assumption $\text{Rad}^{R/A} \leq \mu$ for some maximal torsion radical μ. Since A is $\text{Rad}^{R/A}$-dense, it must be μ-dense, and then A^2 is μ-dense since μ is a torsion radical. But then $\mu \neq 1$ implies that $A^2 \neq 0$, so R is semiprime. If R/A and A are both μ-torsion for a torsion radical μ, then R must be μ-torsion and so $\mu = 1$. Thus $\text{Rad}^{R/A \oplus A} = 1$, since it cannot be

contained in a proper torsion radical, so $R/A \oplus A$ is cofaithful. Hence there exist elements $x_1, \ldots, x_n \in R/A \oplus A$ with $\mathrm{Ann}(x_1, \ldots, x_n) = 0$, where $x_i = y_i + a_i$ for $y_i \in R/A$ and $a_i \in A$. Then $\ell(a_1, \ldots, a_n) \cap A \subseteq \mathrm{Ann}(x_1, \ldots, x_n) = 0$, so $\ell(a_1, \ldots, a_n)\cdot A = 0$. If A is faithful, then $\ell(a_1, \ldots, a_n) = 0$ and A is cofaithful.

(3) ⇒ (5). Let $\{\mu_i\}_{i=1}^n$ be the maximal torsion radicals of R-Mod, with corresponding torsion ideals $\{K_i\}_{i=1}^n$. Then by assumption each μ_i is maximal among proper torsion preradicals, so K_i is a strongly prime ideal by Proposition 1.5. If $K = \cap_{i=1}^n K_i$, then $\mathrm{Rad}^{R/K}(R/K_i) = R/K_i$, so $\mathrm{Rad}^{R/K} \nleq \mu_i$ for all i, and this shows that $\mathrm{Rad}^{R/K} = 1$. It follows that R/K is cofaithful, so K = 0, and R is a subdirect product of finitely many left strongly prime rings.

(5) ⇒ (1). Assume that $\cap_{i=1}^n P_i = 0$ for strongly prime ideals P_i, and that A is a faithful ideal of R. Then for each i, there exist elements $a_{i1}, \ldots, a_{ik} \in A$ such that $ra_{ij} \in P_i$ for all j implies $rA \subseteq P_i$. Then $ra_{ij} = 0$ for all i, j implies $rA \subseteq P_i$ for all i, which implies that $rA \subseteq \cap_{i=1}^n P_i = 0$, so $\ell(\{a_{ij}\}) = 0$ and A is cofaithful. Of course, R is semiprime by assumption. ■

Condition 1 was studied in [6], and condition 3 was given in [3]. Note that R has d.c.c. on left annihilators iff for each right ideal A of R there exist elements $a_1, \ldots, a_n \in A$ such that $\ell(A) = \ell(a_1,\ldots,a_n)$. Condition 4 is essentially Rubin's condition [18] that every proper torsion preradical determines a proper torsion radical, since an argument using Zorn's Lemma can be given to show that in general any proper torsion preradical is contained in a maximal torsion preradical. Handelman [10] showed that conditions 1 and 5 are equivalent. The following corollary is due to Rubin [18].

Corollary 1.7. If R satisfies the conditions of Theorem 1.6, then $Z({}_R R) = 0$ and $Q_{max}(R)$ is isomorphic to a finite direct product of simple von Neumann regular rings.

Proof. If $\{P_i\}_{i=1}^n$ are the minimal prime ideals of R, then their intersection is irredundant. If $x \in R$ and $x \notin P_i$, then $(\cap_{j\neq i} P_j)x \not\subseteq P_i$, and so there exists $r \in \cap_{j\neq i} P_j$ such that $rx \notin P_i$. This shows, as in [10], that as a left R-module, $\oplus_{i=1}^n R/P_i$ is an essential extension of R. It follows that $Z(R) = 0$, since $Z(R/P_i) = 0$ for all i by Proposition 1.4, and that $Q_{max}(R)$ is isomorphic as an R-module to $\oplus_{i=1}^n Q_{max}(R/P_i)$. The isomorphism is easily seen to be a ring homomorphism, and the desired conclusion follows from Proposition 1.4. ∎

Using techniques of free algebras, Handelman and Lawrence [11] have shown that every prime ring can be embedded in a left strongly prime ring, and hence in a simple ring. W. D. Blair (in a private communication) has extended this to show that a semiprime ring can be embedded in a ring satisfying the conditions of Theorem 1.6 iff there exists a finite set of primes $p_1, \ldots, p_s \in \mathbb{Z}$ such that if $0 \neq r \in R$ and $nr = 0$ for some $n \in \mathbb{Z}$, then at least one of the prime $p_1, \ldots, p_s$ divides n.

Finally, we show that the addition of finite uniform dimension to the conditions of Theorem 1.6 gives a characterization of semiprime Goldie rings. Note that condition 3, in particular, gives a characterization which is stated entirely in torsion theoretic language.

Theorem 1.8. [3.6] The following conditions are equivalent for the ring R:

(1) R is a semiprime left Goldie ring;

(2) R is semiprime, every faithful ideal is cofaithful, and every nonzero ideal contains a uniform left ideal;

(3) Every proper torsion preradical of R-Mod is contained in a perfect maximal torsion radical.

Proof. (1) => (2). By assumption, R has d.c.c. on left annihilators and finite uniform dimension.

(2) => (3). If P_i is a minimal prime ideal of R, then the intersection $\cap_{j\neq i} P_j$ over the remaining minimal prime ideals is nonzero,

so it contains a uniform left ideal by assumption. Then R/P_i must contain a uniform left ideal since $\bigcap_{j\neq i} P_j \cap P_i = 0$, so R/P_i has finite uniform dimension since every nonzero left ideal of R/P_i is cofaithful. By [4, Proposition 3.9], the maximal torsion radical determined by $E(R/P_i)$ must be perfect, so the desired conclusion follows from Theorem 1.6.

(3) => (1). By assumption R satisfies the conditions of Theorem 1.6, and by [4, Proposition 3.9] each R-module R/P_i, where P_i is a minimal prime ideal, has finite uniform dimension. Thus R must have finite uniform dimension since it is embedded in a finite direct sum of modules each of which has finite uniform dimension. This completes the proof since R is semiprime and $Z(R) = 0$. ■

§2. Maximal torsion radicals

Proposition 2.1. The following conditions are equivalent for a torsion radical σ of R-Mod:

(1) σ is a maximal torsion radical;

(2) Every nonzero injective object in R-Mod/σ is a cogenerator;

(3) σ is complete and every nonzero torsionfree injective left R_σ-module is faithful;

(4) σ is complete and every nonzero torsionfree injective left $R/\mathrm{rad}_\sigma(R)$-module is faithful;

(5) σ is complete and $\mathrm{rad}_\sigma(R)$ is maximal in the set of proper σ-closed torsion ideals of R.

Proof. Let $\mathrm{rad}_\sigma(R) = K$ and recall that σ is complete iff $\sigma = \mathrm{rad}_{E(R/K)}$.

(1) => (2). Let W be any nonzero injective object in R-Mod/σ. Then W is injective as an R-module, so $\mathrm{rad}_\sigma(W) = 0$ implies that $\sigma \leq \mathrm{rad}_W$. Since σ is maximal, $\sigma = \mathrm{rad}_W$ and W is a cogenerator in R-Mod/σ by [4, Proposition 2.1].

(2) => (3). Since $E(R/K)$ is injective in R-Mod/σ, it must be a cogenerator, so $\sigma = \mathrm{rad}_{E(R/K)}$ and σ is complete. If M is torsionfree and injective as an R_σ-module, then it must be isomorphic to a direct

summand of $\Pi_{\alpha\varepsilon J}E(R_\sigma)$ for some index set J, and since $E(R/K) = E(R_\sigma)$, the latter is injective as an R-module, so M must also be injective as an R-module. Then $\sigma = rad_M$ since $rad_\sigma(M) = 0$, and so M is faithful as as R_σ-module.

(3) => (4). Since $E(R_\sigma) = E(R/K)$, a module is a torsionfree injective R_σ-module iff it is a torsionfree injective R/K-module. If such a module is faithful as an R_σ-module, then it is faithful as an R/K-module.

(4) => (5). If $K \subseteq rad_\tau(R) \neq R$ for a torsion radical τ such that $rad_\tau(R)$ is σ-closed, then $E(R/rad_\tau(R))$ is a torsionfree injective left R/K-module. By assumption $E(R/rad_\tau(R))$ is a faithful R/K-module, so $rad_\tau(R) = K$.

(5) => (1). If $\sigma \leq \tau$ for a proper torsion radical τ, then $rad_\tau(R)$ is σ-closed and $K \subseteq rad_\tau(R)$, so by assumption $K = rad_\tau(R)$. This implies that $\sigma = \tau$ because σ is complete. ∎

Corollary 2.2. Let σ be a maximal torsion radical of R-Mod. Then σ is a prime torsion radical if either R is left Noetherian or σ is perfect.

Proof. Since σ is prime iff R-Mod/σ has an injective cogenerator which is the injective envelope of a simple object in R-Mod/σ, it follows from Proposition 2.1 that σ is prime iff R-Mod/σ has a simple object. If R is left Noetherian, then there exists a maximal σ-closed left ideal A, and $Q_\sigma(R/A)$ is simple in R-Mod/σ. If σ is perfect, then R-Mod/$\sigma \approx R_\sigma$-Mod and must contain simple objects. ∎

Theorem 2.3. [1] Let σ be a perfect torsion radical. Then σ is maximal <=> $R_\sigma/J(R_\sigma)$ is simple Artinian and $J(R_\sigma)$ is right T-nilpotent.

Proof. If σ is perfect, then R-Mod/$\sigma \approx R_\sigma$-Mod, so σ is maximal iff every nonzero injective R_σ-module is a cogenerator. If S is a simple R_σ-module, then E(S) is a cogenerator for R_σ-Mod, and so S must be the only simple R_σ-module (up to isomorphism). For any R_σ-module M, E(M) must be a cogenerator, so E(M) contains an isomorphic copy of S, and this minimal submodule is contained in M since M is essential in E(M). Thus every nonzero injective object of R-Mod/σ is a cogenerator iff R_σ-Mod has only one isomorphism class of simple modules, and every module contains a minimal submodule. This occurs iff $R_\sigma/J(R_\sigma)$ is simple Artinian ($J(R_\sigma)$ denotes the Jacobson radical of R_σ) and $J(R_\sigma)$ is right T-nilpotent, since if $R_\sigma/J(R_\sigma)$ is simple Artinian, then every left R_σ-module contains a minimal submodule iff $J(R_\sigma)$ is right T-nilpotent. (Recall that an ideal A is right T-nilpotent if for each sequence $\{a_i\}_{i=1}^{\infty}$ of elements of A there exists an integer n such that $a_n a_{n-1} \dots a_1 = 0$.) ∎

If the conditions of Theorem 2.3 are satisfied, then R_σ is right perfect, and since there is only one isomorphism class of simple modules, R_σ is isomorphic to the ring of n x n matrices over a local ring. Theorem 2.3 also gives some insight into Goldie's theorem, since if R is a semiprime, left Goldie ring, then $Q_{max}(R)$ is isomorphic to the finite direct product of the rings of quotients defined by the maximal torsion radicals of R-Mod. Each maximal torsion radical μ is perfect with $rad_\mu(R)$ strongly prime, and so R_μ must be simple Artinian since by Proposition 1.4 it is simple and by Theorem 2.3 it is Artinian modulo its Jacobson radical. In general, if R_μ is simple Artinian, then in R-Mod/$\mu \simeq R_\mu$-Mod every nonzero object is a cogenerator, so by Proposition 2.1, μ is maximal. The ring R will be called quasi-local (semi-local) if R/J(R) is simple Artinian (semisimple Artinian).

Corollary 2.4. If R is left Noetherian and σ is a perfect torsion radical then σ is maximal <=> R_σ is a quasi-local left Artinian ring.

Proof. If σ is perfect and R is left Noetherian, then R_σ is left Noetherian. If σ is maximal, then every left R_σ-module contains a minimal submodule, so the ascending socle series of R_σ terminates after a finite number of steps at R_σ. This gives a composition series for R_σ since each term of the series is finitely generated. The converse follows directly from Theorem 2.3. ∎

Corollary 2.5. If R is left hereditary and left Noetherian, then a torsion radical σ is maximal <=> R_σ is simple Artinian.

Proof. Let $rad_\sigma(R) = K$. Since R is hereditary and Noetherian, any homomorphic image of a direct sum of copies of $E({}_{R/K}R/K) = E({}_R R/K)$ is injective as an R-module and hence as an R/K-module. But any direct sum of injective R/K-modules or homomorphic image of an injective R/K-module is of this form, so R/K is left hereditary and left Noetherian. Thus R/K has finite uniform dimension and zero singular ideal, so if σ is maximal, then it is complete and $R_\sigma = Q_{max}(R/K)$ must be semisimple Artinian. Hence R_σ is simple Artinian by Theorem 2.3. The converse follows immediately from Theorem 2.3. ∎

The Walkers [22] showed that the maximal torsion radicals of a commutative Noetherian ring are in one-to-one correspondence with its minimal prime ideals. Theorem 2.8 extends this result to the noncommutative case. The preliminary propositions can be used to show that maximal torsion radicals correspond to minimal prime ideals for any ring with Krull dimension on either the left or right, or for any right perfect ring. The correspondence holds, of course, for any ring which satisfies the conditions of Theorem 1.6, which includes, in the commutative case, any ring with no nilpotent elements and only finitely many minimal prime ideals.

Proposition 2.6. Every proper torsion radical of R-Mod is contained in a maximal torsion radical, and the maximal torsion radicals of R-Mod correspond to the minimal prime ideals of R <=> each nonzero injective left R-module contains a submodule whose left annihilator is a minimal prime ideal.

Proof. =>). If ${}_R W$ is a nonzero injective module, then rad_W is a proper torsion radical, so by assumption $rad_W \leq rad_{E(R/P)}$ for a minimal prime ideal P. Then P must be rad_W-closed, so P is the left annihilator of a submodule of W.

<=). If σ is a proper torsion radical of R-Mod, then $\sigma = rad_W$ for a nonzero injective module W. If P is a minimal prime ideal which is the annihilator of a submodule of W, then $rad_W(R/P) = 0$ shows that $rad_W \leq rad_{E(R/P)}$. If P is any minimal prime ideal of R and $rad_{E(R/P)} \leq \sigma$ for a proper torsion radical σ, then as above, $\sigma \leq rad_{E(R/P')}$ for some minimal prime ideal P'. But then P' is $rad_{E(R/P)}$-closed, so it is the annihilator of a nonzero submodule of E(R/P), which forces $P' \subseteq P$. Since P is minimal we must have P' = P, and thus $\sigma = rad_{E(R/P)}$. It follows that every minimal prime ideal defines a maximal torsion radical of R-Mod, and then every proper torsion radical is contained in a maximal torsion radical. As above, if $rad_{E(R/P)} = rad_{E(R/P')}$, where P and P' are prime ideals, then P = P', and this establishes the one-to-one correspondence between minimal prime ideals of R and maximal torsion radicals of R-Mod. ∎

Proposition 2.7. Every proper torsion radical of R-Mod is contained in a maximal torsion radical, and the maximal torsion radicals of R-Mod correspond to the minimal prime ideals of R <=> R/P(R) satisfies the same condition, for the prime radical P(R) of R, and P(R) is right T-nilpotent.

Proof. Recall that an ideal I is right T-nilpotent iff $\{m \in M \mid Im = 0\} \neq 0$ for every left R-module M.

=>). For any nonzero module $_RM$, by the previous proposition there is a submodule N of E(M) whose left annihilator is a minimal prime ideal P of R. Then $M \cap N \neq 0$ and $P(R) \cdot (M \cap N) = 0$, which shows that P(R) is right T-nilpotent. If M is an injective R/P(R)-module, then $N \subseteq M$, and so by the previous proposition R/P(R) satisfies the desired condition.

<=). If $_RW$ is a nonzero injective R-module, then $M = \{x \in W \mid P(R)x = 0\}$ is nonzero since P(R) is right T-nilpotent. Since M is an injective R/P(R)-module, by assumption there is a submodule of M whose left annihilator in R/P(R) is a minimal prime ideal of R/P(R).

But then the left annihilator in R is a minimal prime ideal of R, and the proof can be completed by applying the previous proposition. ∎

Theorem 2.8 [2]. If R is left Noetherian, then every proper torsion radical of R-Mod is contained in a maximal torsion radical, and the maximal torsion radicals of R-Mod are in one-to-one correspondence with the minimal prime ideals of R.

Proof. If R is left Noetherian, then the prime radical P(R) is nilpotent and R/P(R) is a semiprime left Goldie ring. Since R/P(R) satisfies the conditions of Theorem 1.6, the result follows from Proposition 2.7. ∎

The condition that every proper torsion radical is contained in a maximal torsion radical and that maximal torsion radicals correspond to minimal prime ideals is Morita invariant. A finite direct product of rings satisfies the condition iff each factor does; if a commutative ring has the property, then so does the ring of polynomials over it. A prime ring has the property iff every nonzero injective module is faithful.

In the ring R of linear transformations of an infinite dimensional vector space, the prime ideal P of all linear transformations of finite rank is a torsion ideal since R is von Neumann regular. The next proposition shows that P defines a maximal torsion radical even though P is not a minimal prime ideal of R. The ring C[0, 1] of continuous functions on the interval [0, 1] has no maximal annihilator ideals, and so the same condition holds for its complete ring of quotients Q. It was shown in [1] that the torsion radical determined by E(Q) is not contained in a maximal torsion radical of Q-Mod.

An ideal P which is maximal in the set of proper σ-closed ideals for some torsion radical σ must be a prime ideal. To show this, suppose that $aRb \subseteq P$ with $b \notin P$. If $B = P + RaR$, then $Bb \subseteq P$, so $\mathrm{Hom}_R(R/B, R/P) \neq 0$, which shows that B is not σ-dense. Thus the σ-closure C of B is a proper σ-closed ideal which contains P, a contradiction. Conversely, if P is a prime ideal, $\sigma = \mathrm{rad}_{E(R/P)}$, and C is a proper σ-closed ideal of R, then $C \subseteq P$, so P is a maximal σ-closed ideal. To show this, if $C \nsubseteq P$, then $Cx \neq 0$ for all $0 \neq x \in R/P$ implies that $\mathrm{Hom}_R(R/C, E(R/P) = 0$ and C is σ-dense, a contradiction.

Proposition 2.9. Let σ be a complete torsion radical, with $\mathrm{rad}_\sigma(R) = K$. If K is a prime ideal, then σ is maximal. The converse holds if K is semiprime.

Proof. Since σ is complete, $\sigma = \mathrm{rad}_{E(R/K)}$. If K is prime, then by the above remarks it is a maximal σ-closed ideal of R, so σ is maximal by Proposition 2.1.

Conversely, assume that K is semiprime. By the above remarks, to show that K is prime it is sufficient to show that K is a maximal σ-closed ideal. If A is a proper σ-closed ideal which contains K, then $\mathrm{Hom}_R(K, E(R/A)) = 0$ since K is σ-torsion and $E(R/A)$ is σ-torsionfree. If $0 \neq f \in \mathrm{Hom}_R(A/K, E(R/A))$, then since $\sigma = \mathrm{rad}_{E(R/K)}$, there exists $g \in \mathrm{Hom}_R(E(R/A), E(R/K))$ such that $gf \neq 0$, and consequently there is a left ideal B with $K \subseteq B \subseteq A$ such that $f(B/K) \subseteq R/A$ and $0 \neq gf(B/K) \subseteq R/K$. Let C be the left ideal containing K with $C/K = gf(B/K)$. Then $B \subseteq A$ implies $B \cdot f(B/K) \subseteq B \cdot (R/A) = 0$ and hence $BC \subseteq K$. Since K is a semiprime ideal, this implies that $CB \subseteq K$, and thus $C \cdot gf(B/K) \subseteq gf(C \cdot (B/K)) = 0$, which shows that $C^2 \subseteq K$. Since K is semiprime, this implies that $C \subseteq K$, a contradiction. Thus $\mathrm{Hom}_R(A/K, E(R/A)) = 0$, and so we must have $\mathrm{Hom}_R(A, E(R/A)) = 0$, which shows that $A = \mathrm{rad}_{E(R/A)}(R)$, and so A is a σ-closed torsion ideal of R. By Proposition 2.1, K is a maximal σ-closed torsion ideal, and so A = K. This completes the proof. ∎

Proposition 2.10. Let σ be a complete torsion radical, with $\mathrm{rad}_\sigma(R) = K$. If R_σ is a prime ring, then σ is maximal. Conversely, if σ is maximal, then either $Z({}_{R/K}R/K) = 0$ or $Z({}_{R/K}R/K)$ is essential in R/K, and if $Z({}_{R/K}R/K) = 0$, then R_σ is a prime ring.

Proof. If R_σ is a prime ring, then by Proposition 2.9, $E(R_\sigma)$ defines a maximal torsion radical of R_σ-Mod, and so every nonzero torsionfree injective left R_σ-module is faithful by Proposition 2.1, which then shows that σ is maximal.

Conversely, let σ be a maximal torsion radical. Then $\mathrm{rad}_{E(R/K)}$ defines a maximal torsion radical of R/K-Mod, and since $G \geq \mathrm{rad}_{E(R/K)}$ for the Goldie torsion radical of R/K-Mod, we must have either $G = \mathrm{rad}_{E(R/K)}$, in which case $Z({}_{R/K}R/K) = 0$, or $G = 1$, in which case $Z({}_{R/K}R/K)$ is essential in R/K. If $Z({}_{R/K}R/K) = 0$, then R_σ is von Neumann regular since σ is complete. If A is any proper σ-closed ideal of R_σ, then A is a torsion ideal, so A = 0 by Proposition 2.1, and then the remarks preceding Proposition 2.9 show that R_σ is a prime ring. ∎

§3. Localization at semiprime Goldie ideals

Lambek and Michler [15, Theorem 3.9] showed that if P is a prime ideal of a left Noetherian ring, then the torsion radical $\sigma = \mathrm{rad}_{E(R/P)}$ is defined by a σ-torsionfree indecomposable injective module (which is unique up to isomorphism). Jategaonkar [14] extended this result to prime Goldie ideals. The following theorem gives both necessary and sufficient conditions for a torsion radical to be defined by a prime Goldie ideal, and can thus be viewed as an extension of the module theoretic characterization of prime left Goldie rings given by Faith [8, Theorem 34].

Theorem 3.1. The torsion radical σ is defined by E(R/P) for a prime Goldie ideal P <=> σ is a prime torsion radical defined by a monoform, strongly prime, quasi-injective module which is finitely generated over its endomorphism ring.

Proof. =>). If P is a prime Goldie ideal of R, then let M be a uniform left ideal of R/P. Since R/P is left strongly prime, M is cofaithful as an R/P-module, and monoform and strongly prime as an R-module. The same conditions hold for the quasi-injective envelope $\bar{M}$ of M, which is an R/P-module, and the elements $m_1, \ldots, m_n \in \overline{M}$ such that $\mathrm{Ann}(m_1, \ldots, m_n) = P$ serve as generators for $\overline{M}$ over its endomorphism ring. Now $E(\overline{M}) \subsetneq E(R/P)$, and so $\sigma \leq \mathrm{rad}_{E(\overline{M})}$, while $\mathrm{rad}_{E(\overline{M})} \leq \sigma$ since $\mathrm{Ann}(\overline{M}) = P$.

<=). Let M be a monoform, strongly prime, quasi-injective module which is finitely generated over its endomorphism ring, and which defines σ. If $P = \mathrm{Ann}(M)$, then $R/P \subsetneq M^n$ for some n, since the generators of M over its endomorphism ring can be used to show that M is cofaithful as an R/P-module. Thus R/P is strongly prime and has finite uniform dimension as an R-module, since M is strongly prime and monoform. Since the same conditions hold for R/P as an R/P-module, P is a prime Goldie ideal by Theorem 1.8. The embedding also shows that $\sigma = \mathrm{rad}_{E(M)} \leq \mathrm{rad}_{E(R/P)}$, and that $E({}_{R/P}R/P) \subsetneq M^n$, since M is cofaithful and quasi-injective as an R/P-module and therefore injective as an R/P-module. Because M is cofaithful, it generates $E({}_{R/P}R/P)$, so there exists a nonzero homomorphism $f:M \to E({}_{R/P}R/P)$, and since M is monoform, f followed by the inclusion into M^n must be a monomorphism. Thus f is monic, and so the embedding $f:M \to E(R/P)$ guarantees that $\mathrm{rad}_{E(R/P)} \leq \mathrm{rad}_{E(M)} = \sigma$, which completes the proof. ∎

For a prime ideal P of a commutative ring R, the torsion radical defined by E(R/P) coincides with that defined by the complement of P. In general, for an ideal I of a ring R, let $C(I) = \{c \in R \mid cr \in I \text{ or } rc \in I \text{ implies } r \in I\}$. Thus C(I) denotes the set of elements of R whose images are regular in R/I. The module ${}_RM$ is said to be C(I)-torsion if for each $m \in M$ there exists $c \in C(I)$ such that $cm = 0$, and then a torsion radical $\mathrm{rad}_{C(I)}$ can be defined by letting $\mathrm{rad}_{C(I)}(N)$ be the sum in N of all C(I)-torsion submodules, for all $N \in$ R-Mod. That is, $\mathrm{rad}_{C(I)}(N) = \{x \in N \mid Rx \text{ is } C(I)\text{-torsion}\}$.

To show that $\mathrm{rad}_{C(I)}$ is in fact a torsion radical, we note that if $\mathrm{rad}_{C(I)}(M/\mathrm{rad}_{C(I)}(M)) \neq 0$, then there exists $m \in M$ such that $m \notin \mathrm{rad}_{C(I)}(M)$, but given $r \in R$ there exists $c \in C(I)$ such that

$crm \in rad_{C(I)}(M)$. Then there exists $c' \in C(I)$ such that $c'crm = 0$, and since $c'c \in C(I)$, this implies that $m \in rad_{C(I)}(M)$, a contradiction. The remaining conditions follow easily from the definition since a submodule of a C(I)-torsion module is again C(I)-torsion.

Observe that ${}_RM$ is $rad_{C(I)}$-torsionfree (or simply C(I)-torsionfree) if for each $0 \neq m \in M$ there exists $r \in R$ such that $Ann(rm) \cap C(I) = \emptyset$, and so in particular R/I is C(I)-torsionfree and hence $rad_{E(R/I)} \geq rad_{C(I)}$. If R satisfies the left Ore condition with respect to C(I), that is, if for each $c \in C(I)$ and $r \in R$ there exist $c' \in C(I)$ and $r' \in R$ such that $r'c = c'r$, then for all $M \in R$-Mod, $rad_{C(I)}(M) = \{m \in M \mid cm = 0 \text{ for some } c \in C(I)\}$. Note that this condition holds iff R/Rc is C(I)-torsion for all c C(I), or equivalently, when I is a semiprime Goldie ideal, iff the elements of C(I) are not zero divisors on E(R/I). If R satisfies the left Ore condition with respect to C(0), then the corresponding ring of quotients is the left classical ring of quotients of R, denoted by $Q_{cl}(R)$. If R is a semiprime, left Goldie ring, then $Q_{cl}(R) = Q_{max}(R)$ is semisimple Artinian by Goldie's theorem. The following proposition is due to Jategaonkar [14].

Proposition 3.2. If I is a semiprime Goldie ideal of R, then $rad_{E(R/I)} = rad_{C(I)}$.

Proof. The proof requires only that every dense left ideal of R/I contains a regular element, which includes the case in which R/I is a semiprime left Goldie ring. Since $rad_{C(I)} \leq rad_{E(R/I)}$, suppose that they are not equal. Then there exists a nonzero module ${}_RM$ which is C(I)-torsionfree but $rad_{E(R/I)}$-torsion. By the definition of $rad_{C(I)}$ there exists $m \in M$ such that $A \cap C(I) = \emptyset$ for $A = Ann(m)$, and then $A + I/I$ does not contain a regular element of R/I since $(A + I) \cap C(I) = \emptyset$. Since M is $rad_{E(R/I)}$-torsion, A and (hence) A + I are $rad_{E(R/I)}$-dense in R, and therefore A + I/I is dense in R/I, contradicting the hypothesis. ∎

If I is a semiprime Goldie ideal, and $\{P_i\}_{i=1}^n$ are the minimal prime ideals of R/I, then as in Proposition 1.7, $Q_{cl}(R/I) \simeq \Pi_{i=1}^n Q_{cl}(R/P_i)$. If $c \in C(I)$, then c is invertible in $Q_{cl}(R/I)$, so it must be regular modulo P_i, for each i. The converse can be shown easily, so it follows that $C(I) = \cap_{i=1}^n C(P_i)$. Furthermore, $E({}_{R/I}R/I) \simeq \oplus_{i=1}^n E({}_{R/I}R/P_i)$, and so $E(R/I) \simeq \oplus_{i=1}^n E(R/P_i)$, which implies that $\mathrm{rad}_{E(R/I)} = \cap_{i=1}^n \mathrm{rad}_{E(R/P_i)}$, as was shown by Jategaonkar [14].

Proposition 3.3. Let P and P' be prime Goldie ideals of R. Then $P \subseteq P' \iff \mathrm{rad}_{E(R/P)} \geq \mathrm{rad}_{E(R/P')}$.

Proof. <=). This holds when P' is any prime ideal, by an argument used in both Theorem 1.6 and Proposition 2.6.

=>). This requires only that P is a prime ideal of R for which every nonzero injective R/P-module is faithful, a condition which holds for any strongly prime ideal. Then $P = \mathrm{Ann}(E({}_{R/P}R/P'))$, which shows that P is $\mathrm{rad}_{E(R/P')}$-closed, since $E({}_{R/P}R/P') \subseteq E(R/P')$, and so $\mathrm{rad}_{E(R/P)} \geq \mathrm{rad}_{E(R/P')}$. ∎

For any ideal I of R, the exact sequence $0 \to I \to R \to R/I \to 0$ of R-modules gives rise to the exact sequence $0 \to I_\sigma \to R_\sigma \to (R/I)_\sigma$ of R_σ-modules, since the quotient functor determined by the torsion radical σ is left exact when viewed as a functor from R-Mod into R_σ-Mod. We will identify R_σ/I_σ with the corresponding submodule of $(R/I)_\sigma$. We note that although I is an ideal of R, I_σ may be only a left ideal of R_σ.

Proposition 3.4. Let I be an ideal of R and let $\sigma = \mathrm{rad}_{E(R/I)}$.

(a) $Q_{max}(R/I) = \{x \in (R/I)_\sigma \mid I_\sigma x = 0\}$.

(b) $R_\sigma/I_\sigma \subseteq Q_{max}(R/I) \iff I_\sigma$ is an ideal of R_σ.

(c) If I_σ is an ideal of R_σ, then R_σ/I_σ is a subring of $Q_{max}(R/I)$.

(d) If I_σ is an ideal of R_σ and σ is perfect, then $R_\sigma/I_\sigma = Q_{max}(R/I)$.

Proof. (a) $Q_{max}(R/I)$ is defined by the R/I-injective envelope of R/I, given by $E({}_{R/I}R/I) = \{x \in E(R/I) \mid Ix = 0\}$. Since $E({}_{R/I}R/I)$ is a fully invariant submodule of E(R/I) and $Q_{max}(R) = \{x \in E({}_{R/I}R/I) \mid f(x) = 0$ for all $f \in End(E({}_{R/I}R/I))$ such that $f(R/I) = 0\}$, we must have $Q_{max}(R/I) = E({}_{R/I}R/I) \cap (R/I)_\sigma$. The result follows from the fact that for any element m of a σ-torsionfree left R_σ-module, $Im = 0$ iff $I_\sigma m = 0$.

(b) I_σ is an ideal of R_σ iff $I_\sigma R_\sigma \subseteq I_\sigma$, that is, iff $I_\sigma(R_\sigma/I_\sigma) = 0$, which occurs by part (a) iff $R_\sigma/I_\sigma \subseteq Q_{max}(R/I)$.

(c) If I_σ is an ideal of R_σ, then the R-homomorphism $\pi: R_\sigma \to (R/I)_\sigma$ induced by $R \to R/I \to 0$ maps R_σ into $Q_{max}(R/I)$, and $\pi(1) = 1$. Since for $p, q \in R_\sigma$, $\pi(pq) = \pi\rho_q(p)$ and $\pi(p)\pi(q) = \rho_{\pi(q)}\pi(p)$, where ρ_x defines multiplication by elements of R_σ, to show that π is a ring homomorphism it suffices to show that $\pi\rho_q = \rho_{\pi(q)}\pi$. These map into E(R/I) and so they will be equal if they agree on $R/rad_\sigma(R)$. This holds since $\pi\rho_q(1) = \pi(q) = \rho_{\pi(q)}(1) = \rho_{\pi(q)}\pi(1)$.

(d) If σ is perfect, then Q_σ:R-Mod → R_σ-Mod is exact, so $R_\sigma/I_\sigma = (R/I)_\sigma$, and this implies that $R_\sigma/I_\sigma = Q_{max}(R/I)$. ∎

We note that Proposition 3.4 (a) generalizes Proposition 3.3 of [12], and as in Proposition 3.4 of [12], we can show that for the idealizer $\mathcal{I}(I_\sigma) = \{q \in R_\sigma \mid I_\sigma q \subseteq I_\sigma\}$, we have $\mathcal{I}(I_\sigma)/I_\sigma = Q_{max}(R/I) \cap (R_\sigma/I_\sigma)$. Parts (b) and (c) of Proposition 3.4 extend Lemma 2.3 of [16] and part of Theorem 3.6 of [12]. The next proposition extends parts of Theorems 3.6 and 3.7 of [12].

Proposition 3.5. Let P be a prime ideal of R, with $\sigma = rad_{E(R/P)}$.

(a) P_σ is an ideal of R_σ <=> R_σ/P_σ is a prime R-module.

(b) The following conditions are equivalent:

(1) P_σ is an ideal of R_σ and $(R/P)_\sigma$ is a prime R_σ-module;

(2) $(R/P)_\sigma = Q_{max}(R/P)$;

(3) $(R/P)_\sigma$ is a prime R-module.

Proof. (a) If P_σ is an ideal of R_σ, then $R_\sigma/P_\sigma \subseteq E({}_{R/P}R/P)$, and the latter is prime. Conversely, if R_σ/P_σ is prime, then $P = \mathrm{Ann}(R/P)$ implies $P = \mathrm{Ann}(R_\sigma/P_\sigma)$, so $P_\sigma R_\sigma \subseteq P_\sigma$.

(b) (1) => (2). If P_σ is an ideal of R_σ, then $P_\sigma = \mathrm{Ann}(R_\sigma/P_\sigma)$, so if $(R/P)_\sigma$ is a prime R_σ-module, then $P_\sigma = \mathrm{Ann}((R/P)_\sigma)$ and $(R/P)_\sigma = Q_{max}(R/P)$ by Proposition 3.4.

(2) => (3). $Q_{max}(R/P)$ is a prime R/P-module, so it is prime as an R-module.

(3) => (1). If $(R/P)_\sigma$ is a prime R-module, then P_σ is an ideal by part (a) since $R_\sigma/P_\sigma \subseteq (R/P)_\sigma$. Thus P_σ is a prime ideal, and $(R/P)_\sigma$ is a prime R_σ-module since it is contained in the R_σ/P_σ-injective envelope of R_σ/P_σ. ∎

Theorem 3.6 [7]. Let I be a semiprime Goldie ideal of the ring R. The following conditions are equivalent for $\sigma = \mathrm{rad}_{E(R/I)}$:

(1) $I_\sigma = J(R_\sigma)$ and R_σ is a semi-local ring;

(2) I_σ is an ideal of R_σ and σ is perfect;

(3) R satisfies the left Ore condition with respect to $C(I)$, and for each $c \in C(I)$ there exists $r \in R$ such that $rc \in C(I)$ and such that for each $a \in R$ with $arc = 0$ there exists $c' \in C(I)$ with $c'ar = 0$.

Proof. (1) => (2). If $I_\sigma = J(R_\sigma)$, then $R_\sigma/J(R_\sigma) = R_\sigma/I_\sigma \subseteq E(R/I)$, and if $R_\sigma/J(R_\sigma)$ is semisimple Artinian, then $E(R/I)$ contains an isomorphic copy of each simple left R_σ-module, so it is a cogenerator. It follows from [4, Proposition 3.7] that σ is perfect.

(2) => (1). If I_σ is an ideal and σ is perfect, then by Proposition 3.4, $R_\sigma/I_\sigma = Q_{cl}(R/I)$ is semisimple Artinian since R/I is a semiprime Goldie ring, and thus $I_\sigma \supseteq J(R_\sigma)$. On the other hand, since σ is perfect, $E(R/I) = E(R_\sigma/I_\sigma)$ must contain an isomorphic copy of each simple R_σ-module. Therefore I_σ annihilates each simple R_σ-module, and $I_\sigma \subseteq J(R_\sigma)$.

(1) and (2) => (3). Since R/I is a semiprime left Goldie ring,

$\sigma = \text{rad}_{C(I)}$ and $Rc + I/I$ is σ-dense in R/I for all $c \in C(I)$, so $Rc + I$ is σ-dense for all $c \in C(I)$. Therefore $R_\sigma c + I_\sigma = R_\sigma(Rc + I) = R_\sigma$ since σ is perfect, and since $I_\sigma = J(R_\sigma)$ by condition 1, $R_\sigma = R_\sigma c = R_\sigma Rc$, which shows that Rc is σ-dense. Thus R satisfies the left Ore condition with respect to $C(I)$. The second part of condition 3 holds by [21, Proposition 15.3] since σ is perfect.

(3) => (2). By [21, Proposition 15.3], σ is perfect. Since the left Ore condition is satisfied, R_σ can be constructed as a set of ordered pairs of elements, subject to the usual equivalence relation, and a direct computation shows that $I_\sigma = R_\sigma I$ is an ideal of R_σ. ∎

In Theorem 3.6 the assumption that I is a semiprime Goldie ideal can be removed. In [5] it was shown that for any ideal I of R, $I_\sigma = J(R_\sigma)$ and R_σ is semilocal iff I_σ is an ideal of R_σ, σ is perfect, and the ring R/I has finite uniform dimension and zero singular ideal.

If R is left Noetherian, then condition 3 can be simplified to give Theorem 5.6 of [16]. As Lambek and Michler showed in [16], if R is left Noetherian, I is a semiprime ideal of R, and $ac = 0$ for $a \in R$, $c \in C(I)$, then we can choose a positive integer n such that $\ell(c^n)$ is maximal. If the left Ore condition is satisfied with respect to $C(I)$, then there exist $a' \in R$ and $c' \in C(I)$ such that $a'c^n = c'a$. Since $\ell(c^n)$ is maximal, $\ell(c^{n+1}) = \ell(c^n)$, and then $a'c^{n+1} = c'ac = 0$ shows that $a'c^n = 0$ and hence $c'a = 0$. Thus if R is left Noetherian, then condition 3 of Theorem 3.6 is equivalent to the condition that R satisfies the left Ore condition with respect to $C(I)$. An additional equivalent condition given by Lambek and Michler is that I/Ic is σ-torsion for each $c \in C(I)$.

In the following example from [15], the ring R is left Noetherian but does not satisfy the left Ore condition with respect to $C(P)$ for its prime ideal P, even though the torsion radical defined by $E(R/P)$ is perfect. Let D be a commutative local Noetherian domain such that $J = J(D)$ is principal. Let R be the ring of matrices $\begin{pmatrix} D & J \\ D & D \end{pmatrix}$, and let

P be the prime ideal $\begin{pmatrix} J & J \\ D & D \end{pmatrix}$. The ring of quotients defined by P is just $\begin{pmatrix} D & D \\ D & D \end{pmatrix}$.

On the other hand, any ideal of the full ring of n x n matrices over a commutative ring satisfies condition 3. This can be shown by using the fact that a matrix over a commutative ring is regular iff its determinant is regular.

Corollary 3.7. If the conditions of Theorem 3.6 hold, then (with the notation of Theorem 3.6) the localization M_σ of any R/I-module is an R_σ/I_σ-module, and hence it is a direct sum of simple R_σ-modules.

Proof. If M is an R/I-module, then IM = 0 and hence IM" = 0 for M" = $M/\mathrm{rad}_\sigma(M)$. If $x \in M_\sigma$, then $Dx \subseteq M''$ for some σ-dense left ideal D, and so IDx = 0. But then $I_\sigma Dx = 0$, and so since σ is perfect and I_σ is an ideal we must have $I_\sigma x = I_\sigma R_\sigma x = I_\sigma R_\sigma Dx = I_\sigma Dx = 0$. ■

The following theorem extends Small's theorem [19] characterizing rings whose classical ring of quotients is left Artinian. The Noetherian case was given by Lambek and Michler, although the proof below is taken from [7], in which the theorem was proved in its full generality. The assumption that R is left Noetherian can be dropped if the following conditions are added: (iii) R/I and R/K are left Goldie rings and (iv) $R/\{r \in R \mid I^m r \subseteq K\}$ has finite uniform dimension for all integers $1 \leq m \leq k$, where $I^k \subseteq K$ but $I^{k-1} \not\subseteq K$.

Theorem 3.8 [16]. Let I be a semiprime ideal of a left Noetherian ring R, and let $\sigma = \mathrm{rad}_{E(R/I)}$, with $K = \mathrm{rad}_\sigma(R)$. Then R_σ is a left Artinian classical ring of fractions of R with respect to C(I) <=> (i) $I^k \subseteq K$ for some positive integer k and (ii) for each $r \in R$ and $c \in C(I)$, rc = 0 implies there exists $c' \in C(I)$ such that c'r = 0.

Proof. Assume that R_σ is a left Artinian classical ring of left fractions with respect to C(I). Since R_σ is a partial classical ring of left quotients of R/K, it follows that $R_\sigma = Q_{cl}(R/K)$ since R_σ is left

Artinian. By [21, Proposition 15.7], condition (ii) holds, and thus the conditions of Theorem 3.6 are satisfied, since R must satisfy the left Ore condition with respect to C(I). But then $R_\sigma I = I_\sigma = J(R_\sigma)$ is nilpotent, and so $I^k \subseteq K$ for some positive integer k. This shows that (i) holds and moreover that I/K is the prime radical of R/K.

Conversely, $\mathrm{rad}_{E(R/I)} = \mathrm{rad}_{C(I)}$ by Proposition 3.2, and so each element of C(I) is left regular modulo $K = \mathrm{rad}_{C(I)}(R)$. To show this, let $r \in R$ and $c \in C(I)$ and assume that $rc \in K$. Then for each $s \in R$, there exists $c' \in C(I)$ such that $c'src = 0$, so by condition (ii) there exists $c'' \in C(I)$ such that $c''csr = 0$, and so $r \in K$.

To show that R satisfies the left Ore condition with respect to C(I) it suffices to show that R/Rc is C(I)-torsion for all $c \in C(I)$. Let $I_o = K$ and $I_m = \{r \in R | I^m r \subseteq K\}$, so that $I_k = R$ if $I^k \subseteq K$ but $I^{k+1} \not\subseteq K$. If $rc \in I_m$ for $r \in R$ and $c \in C(I)$, then $I^m rc \subseteq K$, and since c is left regular modulo K, $I^m r \subseteq K$ and hence $r \in I_m$. Thus each element of C(I) is left regular modulo I_m and so $I_m \cap I_{m+1}c = Imc$ for $0 \leq m \leq k$.

Now for $0 \leq m \leq k$, $I_{m+1} \supseteq I_m + I_{m+1}c \supseteq I_{m+1}c$ and $(I_m + I_{m+1}c)/I_{m+1}c \simeq I_m/I_m \cap I_{m+1}c = I_m/I_m c$. Thus to show that $I_{m+1}/I_{m+1}c$ is σ-torsion it suffices to show that both $I_{m+1}/I_m + I_{m+1}c$ and $I_m/I_m c$ are σ-torsion, so that R/Rc is σ-torsion if $I_{m+1}/I_m + I_{m+1}c$ is σ-torsion for $0 \leq m \leq k$ ($I_o/I_o c = K/Kc$ is a factor of $K = \mathrm{rad}_\sigma(R)$ and so it must be σ-torsion).

Let $x \in I_{m+1}$, and let $A = \{r | rx \in I_m + I_{m+1}c\}$. If A/I is essential in R/I, then since R/I is a semiprime Goldie ring, there exists $c' \in C(I)$ such that $c' \in A$ and $I_{m+1}/I_m + I_{m+1}c$ will be σ-torsion. Accordingly, to show that A/I is essential in R/I, let $r \in R$ and $r \notin A$. Then $rx \notin I_m$, since $rx \in I_m$ implies $r \in A$, and so $0 \neq rx + I_m$ and $I_{m+1}c + I_m/I_m$ is essential in I_{m+1}/I_m since c is left regular modulo I_m and R/I_m is by assumption finite dimensional. Thus there exists $s \in R$ such that $srx \in I_{m+1}c + I_m$ but $srx \notin I_m$. Now $sr \in I$ implies

$I^m srx \subseteq I^{m+1}x \subseteq K$, a contradiction. Thus $sr \notin I$, and A/I is essential in R/I.

Since it has now been established that R satisfies the left Ore condition with respect to C(I), if $cr \in K$ for $r \in R$, $c \in C(I)$, then there exists $c' \in C(I)$ such that $c'cr = 0$, so $r \in K$. Thus $C(I) \subseteq C(K)$ and since $I^k \subseteq K$, I/K is the prime radical of R/K, and R/K satisfies the hypothesis of Small's theorem. It follows that C(I) = C(K) and thus $R_\sigma = Q_{cl}(R/K)$ is left Artinian.

References

1. J. A. Beachy, "On maximal torsion radicals," Can. J. Math. 24(1973), 712-726.

2. __________, "On maximal torsion radicals, II," Can. J. Math. 27(1975), 115-120.

3. __________, "On maximal torsion radicals, III," Proc. Amer. Math. Soc., 52(1975), 113-116.

4. __________, "Perfect quotient functors," Comm. in Algebra 2(1974), 403-427.

5. __________, "On localization at an ideal, Canad. Math. Bull. (to appear).

6. __________ and W. D. Blair, "Rings whose faithful left ideals are cofaithful," Pacific J. Math., 58(1975), 1-13.

7. __________,"Localization at semiprime ideals," J. Algebra (to appear).

8. C. Faith, "Modules finite over endomorphism ring," Springer Lecture Notes No. 246 (1972), 145-181.

9. R. Gordon and J. C. Robson, "Krull dimension," Mem. Amer. Math. Soc. 133(1973).

10. D. Handelman, "Strongly semiprime rings," Pacific J. Math. (to appear).

11. __________ and J. Lawrence, "Strongly prime rings," Trans. Amer. Math. Soc. (to appear).

12. A. G. Heinicke, "On the ring of quotients at a prime ideal of a right Noetherian ring," Can. J. Math. 24(1972), 703-712.

13. A. V. Jategaonkar, "Injective modules and localization in non-commutative Noetherian rings," Trans. Amer. Math. Soc. (to appear).

14. __________, "The torsion theory at a semi-prime ideal," Proc. Nat. Acad. Sci. (Brazil).

15. J. Lambek and G. Michler, "The torsion theory at a prime ideal of a right Noetherian ring," J. Algebra 25(1973), 364-389.

16. __________, "Localization of right Noetherian rings at semi-prime ideals," Can. J. Math. 26(1974), 1069-1085.

17. R. A. Rubin, "Absolutely torsion-free rings," Pacific J. Math. 46(1973), 503-514.

18. __________, "Essentially torsion-free rings.

19. L. W. Small, "Orders in artinian rings," J. Algebra 4(1966), 13-41.

20. J. Viola-Prioli, "On absolutely torsion-free rings and kernel functors," Ph.D. Thesis, Rutgers (1973).

21. B. Stenström, "Rings and modules of quotients," Springer Lecture Notes No. 237(1971).

22. C. L. Walker and E. A. Walker, "Quotient categories and rings of quotients," Rocky Mtn. J. Math. 2(1972), 513-555.

ZERO-DIVISORS IN TENSOR PRODUCTS*

George M. Bergman
University of California
Berkeley, California

I shall discuss a very simple, frustrating, tantalizing question in ring theory, some partial results I have obtained on it, and some possible methods of further attack, which the reader is invited to pursue. The problem comes from the theory of division algebras, but the approach to be described will lead us through general ring theory, linear algebra, algebraic geometry and category theory. (We shall make a detour into semigroup theory as well.)

Which fields may hold the methods by which the problem can be solved I do not know; but I think this paper contains ideas which should be of interest to people in all the areas mentioned.

THE MAIN THREAD

§1. The Problem.

Fields are commutative, and rings algebras and semigroups will all have 1.

At the 1971 Park City Conference in Ring Theory, Claudio Procesi (in conversation) asked:

(1) Let K be an algebraically closed field, and D, E division algebras over K which are finite-dimensional over their respective centers. Will $D \otimes_K E$ be a K-algebra without zero-divisors?

*Much of this work was done while the author was partly supported by NSF contract GP 9152, and some of the preparation of the manuscript took place while he was supported by NSF contract MPS 73-08528.

MOS Classifications: Primary: 15 A 69, 16 A 02, 16 A 40.
Secondary: 14 N 05, 16 A 34, 16 A 49, 17 E 05, 18 B 99, 18 D 99.

Let me recall some examples of how zero-divisors <u>are</u> known to arise in tensor products of algebras not having them. The first of these is well-known; but we will refer to it again, so we record it as:

<u>Lemma 1.1</u>. Let K be any field, L and L' extensions fields of K, and suppose L simple algebraic, $L = K(\alpha)$. Then $L \otimes L'$ has zero-divisors if and only if the minimal polynomial of α, $P \in K[t]$, factors non-trivially in $L'[t]$. (E.g., if $L = L' \neq K$.) ∎

A variation on this example: Start with two field extensions L, L' of K, whose tensor product has zero-divisors as above, and form the polynomial rings $L[X]$, $L'[Y]$. Now let $R \subseteq L[X]$, $R' \subseteq L'[Y]$ denote the K-subalgebras of all polynomials with <u>constant term</u> in K. Clearly, R and R' have no elements algebraic over K. Yet their tensor product has zero-divisors: if u, v are zero-divisors of $L \otimes_K L'$, then $u\,X\,Y$, $v\,X\,Y$ are zero-divisors in $R \otimes R' \subseteq L[X] \otimes L'[Y]$.

Another sort of example: Let $\mathbb{R}$ and $\mathbb{H}$ denote the real numbers and the quaternions, and L the field of fractions of the commutative $\mathbb{R}$-algebra $\mathbb{R}[X, Y]/(1 + X^2 + Y^2)$. Then $\mathbb{H} \otimes_{\mathbb{R}} L$, i.e., the quaternions with scalars extended to L, <u>has</u> zero-divisors: $(1 + Xi + Yj)(1 - Xi - Yj) = 0$, although L is a field in which $\mathbb{R}$ is algebraically closed. But, of course, $\mathbb{R}$ is <u>not</u> algebraically closed in $\mathbb{H}$. And indeed the existence of nontrivial algebraic extensions of K is clearly involved, directly or indirectly, in each of these examples.

Thus, the algebraic closure assumption in question (1) gives us hope for an affirmative answer!

However, I know very little of the theory of finite-dimensional division algebras, so I thought I would look at the more general question:

(2) Let K be an algebraically closed field and D, E arbitrary division algebras over K. Will the K-algebra $D \otimes_K E$ be without zero-divisors?

This is also the form in which Murray Schacher posed the problem at the Park City problem session [1, p. 380, problem 26].

But I'm really not an expert on any kind of division algebras. So —

(3) Let K be an algebraically closed field, and let R, S be associative K-algebras without zero-divisors. Will $R \otimes_K S$ be without zero-divisors?

After pondering this question, I decided that the first thing that needed to be looked at was the basic linear algebra involved in the construction of the multiplication operation in a tensor-product of algebras. For this purpose, let us define a <u>bilinear system</u> over a field K to mean a 4-tuple $A = (A_1, A_2, A_3, \phi_A)$, where A_1, A_2, A_3 are K-vector-spaces, and $\phi_A\colon A_1 \times A_2 \to A_3$ is a bilinear map. If A, B are two bilinear systems, we can form a bilinear system $A \otimes B = (A_1 \otimes B_1, A_2 \otimes B_2, A_3 \otimes B_3, \phi_A \otimes \phi_B)$, where

$$(\phi_A \otimes \phi_B)(a \otimes b, x \otimes y) =_{\text{def.}} \phi_A(a,x) \otimes \phi_B(b,y)$$

$$(a \in A_1,\ x \in A_2,\ b \in B_1,\ y \in B_2).$$

Note that every K-algebra R has an "underlying" bilinear system $\underline{U}(R)$, defined by $\underline{U}(R)_1 = \underline{U}(R)_2 = \underline{U}(R)_3 = |R|$ (as K-vector-space), $\phi_{\underline{U}(R)}$ = multiplication in R. We see from our definition of tensor product of bilinear systems that $\underline{U}(R \otimes_K S) = \underline{U}(R) \otimes_K \underline{U}(S)$.

Let us define "zero-divisors" in a bilinear system A to mean elements $a \in A_1$, $x \in A_2$ such that $\phi_A(a, x) = 0$ in A_3. Zero-divisors a, x will be called "proper" if a and x are both nonzero; the phrase "without zero-divisors" will be understood to mean "without proper zero-divisors". Then question (3) is generalized by:

(4) Let K be an algebraically closed field, and A, B be bilinear systems over K, without zero-

divisors. Will be bilinear system $A \otimes_K B$ be without zero-divisors?

§2. An equivalence

The reader may feel that in going from question (3) to question (4) we have abandoned all that is beautiful and useful.

To determine how far afield we have actually gone, let us consider a natural construction going from bilinear systems back to associative algebras. Given a bilinear system A over any field K, form the tensor algebra $K < A_1 \oplus A_2 \oplus A_3 >$, and divide out by the ideal thereof generated by $\{a\,x - \phi_A(a, x) \mid a \in A_1,\ x \in A_2\}$. The resulting K-algebra, which will be denoted $K < A >$, can be characterized as <u>universal</u> among K-algebras R given with a homomorphism of A into their multiplicative structure, $f{:}A \to \underline{U}(R)$. (A homomorphism of bilinear systems, $f{:}A \to B$, means a 3-tuple of linear maps, $f_i{:}A_i \to B_i\ (i=1,2,3)$ such that $f_3(\phi_A(a, x)) = \phi_B(f_1(A), f_2(x))$ for $a \in A_1$, $x \in A_2$).

Let X_1, X_2, X_3 be bases for the K-vector-spaces A_1, A_2, A_3, and let X denote their disjoint union $X_1 \amalg X_2 \amalg X_3$. Then it is easy to show that a basis for $K\langle A\rangle$ is given by the set of all monomials w in the elements of X (including the empty monomial 1) such that in w, no element of X_1 is immediately followed by an element of X_2. Note that X_1, X_2 and X_3 will be contained in this basis. Hence the map $A \to \underline{U}(K\langle A\rangle)$ is an embedding, and if A has zero-divisors so does $K\langle A\rangle$.

Now if we grade the tensor algebra $K < A_1 \oplus A_2 \oplus A_3>$ by the free semigroup (with 1) S on two generators p and q, by making A_1 of degree p, A_2 of degree q, and A_3 of degree pq, our relations defining $K\langle A\rangle$ are homogeneous, and $K\langle A\rangle$ inherits an S-grading. From the above description of a basis of $K\langle A\rangle$ it follows that each homogeneous component will be a tensor-product space, $A_{i_1} \otimes \ldots \otimes A_{i_r}$, where the indices $i_1, \ldots, i_r \in \{1, 2, 3\}$ are subject to the condition that there be no two successive terms $i_t = 1$, $i_{t+1} = 2$, (e.g., the component of degree pqp is not $A_1 \otimes A_2 \otimes A_1$ but $A_3 \otimes A_1$.) Since S is an orderable semigroup with cancellation, K<A> will have zero-divisors if and only if it has homogeneous zero-divisors. We see that the only place homogeneous zero - divisors can possibly occur is in one of the multiplications

$$(\ldots \otimes A_1) \times (A_2 \otimes \ldots) \rightarrow (\ldots \otimes A_3 \otimes \ldots).$$

But each of these maps is induced by ϕ_A on the middle factors, and it follows easily that no zero-divisors arise unless A itself has them. Thus we get:

Lemma 2.1. Let A be a bilinear system over an arbitrary field K. Then K<A> has zero-divisors if and only if A does. ∎

It follows that the generalization from (3) to (4) was not drastic at all.

Corollary 2.2. Questions (3) and (4) above are equivalent.

Proof. We already know that a negative answer to (3) would give a negative answer to (4). Conversely, suppose A and B are bilinear systems over K without zero-divisors such that $A \otimes B$ has zero-divisors. By Lemma 2.1, K<A> and K<B> are K-algebras without zero-divisors, which contain embedded images of A and B. Thus K<A> ⊗ K<B> contains an embedded image of $A \otimes B$, and so has zero-divisors, giving a counter-example to (3). ∎

(Remark: K<A ⊗ B> ≇ K<A> ⊗ K<B>. The former ring in general embeds properly in the latter.)

§3. Duality.

An advantage of studying bilinear systems is that, unlike K-algebras, they are obviously always the directed unions of their finite-dimensional subsystems (subsystems whose components A_1, A_2, A_3

are each finite-dimensional). Hence question (4) reduces to the case where A and B are finite-dimensional--a question in basic linear algebra! To this situation, we can apply a sort of "method of undetermined coefficients":

Let us fix a finite-dimensional bilinear system A, over an arbitrary field K. Choose bases $\{a_1, \ldots, a_m\}$ for A_1, $\{x_1, \ldots, x_n\}$ for A_2, and $\{p_1, \ldots, p_r\}$ for A_3, and write the mn products $\phi_A(a_i, x_j) \in A_3$ as linear combinations of the p_k's. We now construct another bilinear system B as follows. B_1 will be a vector space of dimension m, with a basis $\{b_1, \ldots, b_m\}$, and B_2 a space of dimension n, with basis $\{y_1, \ldots, y_n\}$. Form the two elements

$$s = a_1 \otimes b_1 + \ldots + a_m \otimes b_m \in A_1 \otimes B_1,$$

$$t = x_1 \otimes y_1 + \ldots + x_n \otimes y_n \in A_2 \otimes B_2,$$

and take for B_3 the quotient-space of $B_1 \otimes B_2$ by precisely those relations needed to make $\phi_{A\otimes B}(s, t) \in A_3 \otimes B_3$ vanish. To see how to find these relations, let B' be the system $(B_1, B_2, B_1 \otimes B_2, \otimes)$. Let $u = \phi_{A\otimes B'}(s,t) \in A_3 \otimes B_3'$, write u as $p_1 \otimes q_1 + \ldots + p_r \otimes q_r$ $(q_i \in B_3')$ and take $B_3 = B_3'/(q_1, \ldots, q_r)$.

<u>Lemma 3.1</u>. If A is a finite-dimensional bilinear system, then the above construction gives a finite-dimensional bilinear system B whose tensor product with A has zero-divisors $s \in (A \otimes B)_1$, $t \in (A \otimes B)_2$, which are <u>universal</u>, in the sense that for any other bilinear system C (finite- or infinite-dimensional), and zero-divisors s', t' (proper or not) in $A \otimes C$, there is a unique homomorphism of bilinear systems $f:B \to C$ such that $id_A \otimes f: A \otimes B \to A \otimes C$ carries s, t to s', t'. ∎

To understand this construction better, let us go back to the subconstruction implicit therein, in which we associated to a finite-dimensional vector-space U (A_1 or A_2 above) a vector-space $\bar{U}$ (B_1 or B_2) with a universal tensor $s_U \in U \otimes \bar{U}$. There seemed to be nothing to this--we chose any $\overline{U}$ of the same dimension as U, and in effect any tensor s_U of maximal rank in $U \otimes \overline{U}$. Which means that we really didn't understand what we were doing. Note that when we make such choices for every U, the construction becomes functorial. A homomorphism of finite-dimensional vector-spaces, $f:U \rightarrow V$ induces a homomorphism $f \otimes id_U : U \otimes \overline{U} \rightarrow V \otimes \overline{U}$, which carries s_U to some tensor $s' \rightarrow V \otimes \overline{U}$; and by the universal property of s_V, there will be a unique $\overline{f}:\overline{V} \rightarrow \overline{U}$ carrying s_V to s'.

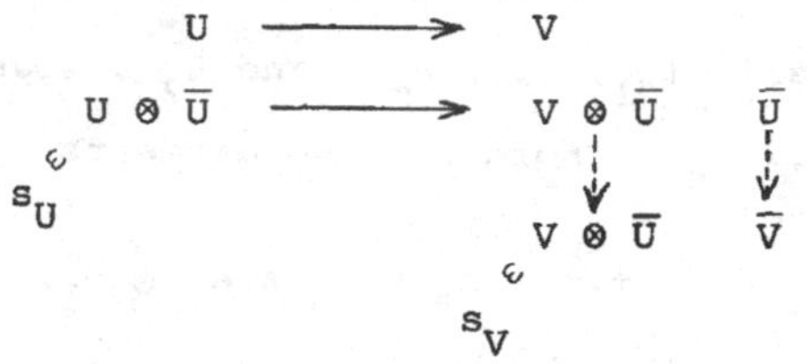

It is easy to check that in this way the construction "-" becomes a contravariant functor on finite-dimensional K-vector-spaces. To end the mystery, we give:

<u>Lemma 3.2</u>. The contravariant functor "-" on <u>finite-dimensional</u> K-vector-spaces is (up to natural isomorphism) the duality functor $(\ )^* = \mathrm{Hom}(-, K)$. For each space U, the element s_U corresponds to $id_U \in U \otimes U^* \cong \mathrm{Hom}(U, U)$. If $u_1, \ldots, u_m$ is a basis for U, and we write $s_U = \sum u_i \otimes \overline{u}_i \in U \otimes U^*$ then $\{\overline{u}_i\}$ is the basis of U^* dual to the basis $\{u_i\}$ of U.

<u>Proof</u>. The functor "-" is determined up to natural isomorphism by its universal property, and "*" (with the elements $id_U \in U \otimes U^*$) is easily shown to have this property. ■

So in our construction on bilinear systems the spaces B_1, B_2 should be looked at as A_1^*, A_2^*. To interpret B_3, recall that the bilinear map $\phi_A : A_1 \times A_2 \rightarrow A_3$ corresponds to a linear map

$\phi'_A : A_1 \otimes A_2 \to A_3$. It is easily verified from its construction that B_3 is the cokernel of the dual map $(\phi'_A)^* : A_3^* \to (A_1 \otimes A_2)^* \cong A_1^* \otimes A_2^*$. Equivalently, if we let A_0 denote the kernel of ϕ'_A:

$$(5) \qquad 0 \longrightarrow A_0 \longrightarrow A_1 \otimes A_2 \xrightarrow{\phi'_A} A_3$$

then $B_3 = A_0^*$. Assume that ϕ'_A is surjective (clearly no loss of generality). Then (5) becomes a short exact sequence:

$$(6) \qquad 0 \longrightarrow A_0 \longrightarrow A_1 \otimes A_2 \longrightarrow A_3 \longrightarrow 0.$$

and B is the bilinear system corresponding to the dual sequence

$$(7) \qquad 0 \longrightarrow A_3^* \longrightarrow A_1^* \otimes A_2^* \longrightarrow A_0^* \longrightarrow 0.$$

Hence let us (re)name this bilinear system A^*. (Caveat: $(A^*)_i = A_i^*$ for i = 1, 2, but $(A^*)_3 = A_0^*$.)

To investigate now whether the tensor product of our system A with a bilinear system C without zero-divisors can have zero-divisors, it is natural to begin by asking whether the universal system A^* is a system "C" without zero-divisors.

The latter question is not equivalent to the former for a given A, since even if A^* has zero-divisors, it may have a non-degenerate homomorphism f into a system C without zero-divisors, which would lead to proper zero-divisors in $A \otimes C$. ("Non-degenerate" meaning f_1 and f_2 both nonzero.) But note that any homomorphic image of A^* can be written A'^*, where A' is (loosely stated) "A with possibly some generators deleted from A_1 and A_2, and then some relations deleted from A_3." Such an A' will be without zero-divisors if A is. Hence if there exist systems without zero-divisors, A, C, such that $A \otimes C$ has zero-divisors, there must exist systems of the form A' and $C' = A'^*$ with these properties. It follows that question (4) has a <u>negative</u> answer if and only if there exists a non-degenerate bilinear system A (i.e., A_1 and A_2 nonzero) such that <u>neither</u> A nor A^* has zero-divisors.

§4. The categories $\underline{C}$ and $\underline{C}_{inf}$.

In this section we pause to formalize in category-theoretic language some of the ideas introduced above, in the process readjusting our definition of a bilinear system to take into account the symmetry between A_3 (which appeared in our original definition) and A_0 (which did not), and making some further observations and definitions.

We fix a field K, and define a category $\underline{C}$ as follows: The objects of $\underline{C}$ will be 6-tuples $A = (A_0, A_1, A_2, A_3; \psi_A, \phi_A)$, where the A_i are finite-dimensional vector-spaces over K, and $\psi_A\colon A_0 \to A_1 \otimes A_2$, $\phi_A\colon A_1 \otimes A_2 \to A_3$ are linear maps making a short exact sequence (6). We shall henceforth call these 6-tuples A "bilinear systems". A_i will be called the i-component of A. A morphism $f:A \to B$ in $\underline{C}$ will be a 4-tuple (f_0, f_1, f_2, f_3) of linear maps $f_i:A_i \to B_i$ which induces a commuting diagram of short exact sequences:

$$\begin{array}{ccccccccc} 0 & \longrightarrow & A_0 & \longrightarrow & A_1 \otimes A_2 & \longrightarrow & A_3 & \longrightarrow & 0 \\ & & \downarrow{\scriptstyle f_0} & & \downarrow{\scriptstyle f_1 \otimes f_2} & & \downarrow{\scriptstyle f_3} & & \\ 0 & \longrightarrow & B_0 & \longrightarrow & B_1 \otimes B_2 & \longrightarrow & B_3 & \longrightarrow & 0. \end{array}$$

Let A and B be bilinear systems and $f_1:A_1 \to B_1$, $f_2:A_2 \to B_2$ linear maps. Note that there will be <u>at most one</u> choice of f_0 and f_3 such that (f_0, f_1, f_2, f_3) is a morphism $f:A \to B$, and that if there in fact exists either f_0 or f_3 making a commutative diagram with $f_1 \otimes f_2$, then the other also exists and we have our f.

It is frequently convenient to specify a bilinear system by showing the exact sequence (6). But there is the difficulty that if A_1 and/or A_2 is itself a tensor product of other vector spaces, the intended factors A_1 and A_2 of the middle term may not be clear. Hence let us make the convention of writing (6) as:

$$(8) \qquad 0 \longrightarrow A_0 \longrightarrow \langle A_1, A_2\rangle \longrightarrow A_3 \longrightarrow 0$$

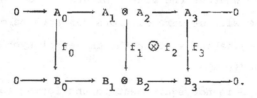

Note that a bilinear system (like a morphism of such systems) is essentially determined by just the right-hand or the left-hand two-thirds of (8). When we wish to specify a system by just giving (say) the former information, we shall simply show:

$$(9)\quad 0 \longrightarrow \cdot \longrightarrow \langle A_1, A_2\rangle \longrightarrow A_3 \longrightarrow 0,$$ (e.g., (10), (12) below).

The construction $(\ )^*$ of (7) clearly describes a duality in $\underline{C}$; a contravariant endofunctor whose square is canonically isomorphic to the identity.

Given objects A and B of $\underline{C}$, let $A \otimes B$ be the object

$$(10)\qquad 0 \longrightarrow \cdot \longrightarrow \langle A_1 \otimes B_1, A_2 \otimes B_2\rangle \longrightarrow A_3 \otimes B_3 \longrightarrow 0.$$

"$\otimes$" is clearly a functor $\underline{C} \times \underline{C} \to \underline{C}$. Note that the left-hand term in (10) is not $A_0 \otimes B_0$! To describe this term, let us abbreviate $A_1 \otimes A_2$ to A_{12} and $B_1 \otimes B_2$ to B_{12}, and assume $A_0 \subseteq A_{12}$, $B_0 \subseteq B_{12}$. Then writing the middle vector space of (10) as $A_{12} \otimes B_{12}$, the kernel term is the subspace $A_0 \otimes B_{12} + A_{12} \otimes B_0$. (This is analogous to the observation that if M and M' are modules over a commutative ring presented by generators g_i (respectively $g_{i'}'$) and relations r_p (respectively $r_{p'}'$) and we represent $M \otimes M'$ in terms of the generators $g_i \otimes g_{i'}'$, then defining relations will be given, not by $\{r_p \otimes r_{p'}'\}$, but by $\{r_p \otimes g_{i'}', g_i \otimes r_{p'}'\}$.)

In view of this asymmetry in (10), $(A \otimes B)^*$ will not be isomorphic to $A^* \otimes B^*$; rather

$$(11)\qquad (A \otimes B)^* \cong A^* \otimes^* B^*,$$

where $\otimes^*: \underline{C} \times \underline{C} \to \underline{C}$ is the functor taking A, B in $\underline{C}$ to the bilinear system

$$(12)\qquad 0 \to A_0 \otimes B_0 \to \langle A_1 \otimes B_1, A_2 \otimes B_2\rangle \to \cdot \to 0.$$

<u>Exercise 4.1</u>. There is a morphism of functors $\otimes^* \to \otimes$ which is the identity map on the 1- and 2-components. ∎

We see that the bifunctors $\otimes$ and $\otimes^*$ are each associative. In §17 we shall examine the relations among $(A \otimes B) \otimes^* C$, $A \otimes (B \otimes^* C)$, etc.

Let us define the object $\underline{1}$ of $\underline{C}$ to be $\underline{U}(K)$, i.e.,

$$0 \to 0 \to \langle K, K\rangle \to K \to 0.$$

Clearly $\underline{1}$ is an identity object for the operation $\otimes$; and the dual object, $\underline{1}^*$:

$$0 \to K \to \langle K, K\rangle \to 0 \to 0$$

is likewise an identity for $\otimes^*$. Note that the set of pairs of zero-divisors in a bilinear system A can be identified with $\mathrm{Hom}(\underline{1}^*, A)$! In particular, proper zero-divisors correspond to non-degenerate maps $\underline{1}^* \to A$; where we define a morphism $f:A \to B$ in $\underline{C}$ to be <u>degenerate</u> if f_1 <u>or</u> f_2 is the zero map; equivalently, if f_0 and f_3 are <u>both</u> zero.

It is easy to verify that the monomorphisms of $\underline{C}$ are the morphisms f such that f_1 and f_2 are 1-1. (Sufficiency is clear. One gets necessity using the test-object $\underline{1}$.) When this is so f_0 will also be 1-1, but f_3 may not be--consider, for instance, the obvious map $\underline{1} \to \underline{1}^*$. If $f:A \to B$ is 1-1 in <u>all four</u> components, we shall say A is (up to isomorphism) a <u>subobject</u> of B.

Dually, f is an epimorphism if and only if f_1 and f_2 are onto. In this case, f_3 will also be onto, but not necessarily f_0.

An arbitrary map $f:A \to B$ in $\underline{C}$ has a natural factorization $A \overset{\alpha}{\to} E \overset{\beta}{\to} F \overset{\gamma}{\to} B$, where

$$E \text{ is } 0 \to f(A_0) \to \langle f(A_1), f(A_2)\rangle \to \cdot \to 0, \quad \text{and}$$

$$F \text{ is } 0 \to \cdot \to \langle f(A_1), f(A_2)\rangle \to f(A_3) \to 0$$

The map α in this factorization will be surjective in all components, the map β will be both epic and monic (like $\underline{1} \to \underline{1}^*$), and γ will be the inclusion of a subobject. We see that either E or F is a reasonable candidate to be called the "image" of f. In deference to our original viewpoint which emphasized A_3 rather than A_0, we will use this term for F. Thus, homomorphic images of A are precisely the epimorphs of A. Their duals, systems monomorphically mapped to A^*, form a larger class than the subobjects. (Cf. last paragraph of §3, where A' has this relation to A.) Images of A under maps all of whose components are surjective will be called "precise homomorphic images." We won't refer very often to the terminology of this paragraph, but I felt these distinctions should be made to avoid confusion.

$\underline{C}$ is a full subcategory of the larger category $\underline{C}_{inf}$ of possibly-infinite-dimensional bilinear systems. All the definitions and observations we have made for $\underline{C}$ go over to $\underline{C}_{inf}$, except for the existence and properties of *. (We even have $\otimes^*$, defined by (12); but of course (11) makes no sense.) Every object of $\underline{C}_{inf}$ is a directed union of subobjects which belong to $\underline{C}$. In any given context we will make clear whether "bilinear system" means object of $\underline{C}$ or of $\underline{C}_{inf}$.

The underlying-multiplicative-system construction $\underline{U}$ defined in §1 is a functor K-alg $\to \underline{C}_{inf}$.

§5. General position arguments.

Let us now think of a finite-dimensional bilinear system A as determined by two finite-dimensional vector-spaces A_1 and A_2, and a subspace $A_0 \subseteq A_1 \otimes A_2$. (So $A_3 = A_1 \otimes A_2/A_0$.) Let $\underline{R} \subseteq A_1 \otimes A_2$ denote the closed subvariety (in the sense of algebraic geometry) of reducible (rank ≤ 1) tensors $a \otimes x$. Then A will have zero-divisors if and only if the subspace A_0 has nontrivial intersection with $\underline{R}$.

This suggests that for K algebraically closed, we can get information about zero-divisors by looking at the dimensions of $\underline{R}$ and A_0.

Let A_1, A_2 and A_0 have dimensions $m > 0$, $n > 0$ and $mn-p$ (so p is the dimension of A_3). The set $\underline{R}$ will have dimension $m + n - 1$. (The "-1" is because a non-zero reducible tensor $a \otimes x$ does not quite determine the pair (a, x): $a \otimes x = (\lambda a) \otimes (\lambda^{-1}x)$ for all $\lambda \in K - \{0\}$ so one degree of freedom is lost.) The subvarieties A_0 and $\underline{R}$ of $A_1 \otimes A_2$ are closed under multiplication by scalars, and so correspond to subvarieties of the $(mn-1)$-dimensional <u>projective space</u> associated to $A_1 \otimes A_2$. Since projective space is complete, A_0 and $\underline{R}$ must have nontrivial intersection if the dimensions of these projective varieties add up to at least the dimension of the ambient projective space, i.e., if $(m + n - 2) + (mn - p - 1) \geq mn - 1$.

There is an almost-converse: if we start with vector spaces A_1 and A_2 of dimensions m, n and let p be an integer such that

$$(13) \qquad (m + n - 2) + (mn - p - 1) < mn - 1.$$

then an $(mn-p)$-dimensional subspace $A_0 \subseteq A_1 \otimes A_2$ in <u>general position</u> will <u>not</u> meet the variety of reducible tensor except at 0. "In general position" means outside of some specifiable proper closed subvariety of the Grassmann variety of p-dimensional subspaces of $A_1 \otimes A_2$. Simplifying (13) (in particular, transforming it into a sharp inequality (14)) these observations become:

<u>Lemma 5.1</u>. Let K be an algebraically closed field, A_1 and A_2 K-vector-spaces of positive finite dimensions m and n respectively. Suppose we take a bilinear system

$$0 \to A_0 \to \langle A_1, A_2\rangle \to A_3 \to 0,$$

and let $p = \dim A_3$, $mn-p = \dim A_0$. Then a necessary condition for A to be without zero-divisors is

$$(14) \qquad p \geq m + n - 1.$$

If A_0 is taken in general position, this is also sufficient. ∎

We can also apply Lemma 5.1 to A^*. Since $(A^*)_3 = (A_0)^*$, which has dimension mn-p, the "necessary and in general sufficient" condition for A^* to be without zero-divisors is

(15) $\quad mn-p \geq m + n - 1.$

To solve (14) and (15) simultaneously, let us add them; the result simplifies to

(16) $\quad (m - 2)(n - 2) \geq 2.$

For all m and n satisfying (16) we see that there will exist p satisfying (14) and (15). Taking the smallest solution of (14)-(16) which is (up to interchange of m and n) $m = 4$, $n = 3$, $p = 6$, we have by the observation at the end of §3:

<u>Corollary 5.2</u>. The answer to question (4), and hence also to (3) is negative. Specifically there is a counterexample to (4) in which $\dim A_1 = 4$, $\dim A_2 = 3$, $\dim A_0 = \dim A_3 = 6$, and $B = A^*$, and thus there is a counterexample to (3) in which R and S are each algebras presented by 7 generators and 6 relations.

§6. An explicit example.

Once I knew what dimensions to look in, it was not hard to experiment and find an example. An explicit look at this example will cheer us up after the above disappointment.

Let K be any field and λ be any element of $K-\{0\}$, and define a bilinear system $S(\lambda)$ as follows. $S(\lambda)_1$ will be a 4-dimensional vector space on a basis $\{a, b, c, d\}$, and $S(\lambda)_2$ 3-dimensional on a basis $\{x, y, z\}$. The space $S(\lambda)_3$ will be 6-dimensional, presented by 12 generators $a\cdot x, \ldots, d\cdot z$ and 6 relations

(17) $\quad a\cdot x = c\cdot y \qquad a\cdot y = c\cdot z \qquad a\cdot z = d\cdot x$

$\qquad\quad b\cdot x = d\cdot y \qquad b\cdot y = \lambda d\cdot z \qquad b\cdot z = c\cdot x.$

Note that the 6 elements represented by the left-hand sides of the above equations will form a basis of $S(\lambda)_3$. The map $\phi_{S(\lambda)}$ is, of course, defined to take $a \otimes x$ to $a \cdot x$, etc.

Lemma 6.1. For $\lambda \in K - \{0\}$, $S(\lambda)^* \cong S(\lambda^{-1})$.

Proof. For bases of $(S(\lambda)^*)_1$ and $(S(\lambda)^*)_2$ let us take the dual bases $\{a^*, \ldots\}$ and $\{w^*, \ldots\}$ to the given bases of $S(\lambda)_1$ and $S(\lambda)_2$. Then $(S(\lambda)^*)_3$ will be spanned by the 12 elements $a^* \cdot x^*, \ldots, d^* \cdot z^*$, which are subject to 6 relations, induced by any basis of $(S(\lambda)^*)_0 = (S(\lambda)_3)^*$. Using the dual basis of the basis of $S(\lambda)_3$ mentioned above, we find that these relations are:

$$(18) \quad a^* \cdot x^* + c^* \cdot y^* = 0 \qquad a^* \cdot y^* + c^* \cdot z^* = 0 \qquad a^* \cdot z^* + d^* \cdot x^* = 0$$

$$b^* \cdot x^* + d^* \cdot y^* = 0 \qquad b^* \cdot y^* + \lambda^{-1} d^* \cdot z^* = 0 \qquad b^* \cdot z^* + c^* \cdot x^* = 0.$$

In terms of the generators $a^*, b^*, -c^*, -d^*; x^*, y^*, z^*$, this system of relations has precisely the form (17), with λ^{-1} in place of λ, as desired. (The reader might also like to derive (18) himself by an "undetermined coefficients" computation, as indicated at the beginning of §3.) ∎

To study the question of whether $S(\lambda)$ has zero-divisors, we will need the following Lemma. Recall that a morphism f of bilinear systems is called degenerate if f_1 or f_2 is 0.

Lemma 6.2. If $\lambda \in K - \{0, 1\}$, and D is any division algebra over K, then any homomorphism of $S(\lambda)$ into $\underline{U}(D)$ is degenerate.

Proof. We first observe that for elements a, b, c, d, x, y, z of any group, the three equations $ax = cy$, $bx = dy$, $ay = cz$ imply $by = dz$. (Solve the second for x, substitute into the first, cancel y, substitute into the third, and the result is equivalent to the last). Hence under any homomorphism $f: S(\lambda) \to \underline{U}(D)$, if all seven basis elements went into the group of units of D, we would get a contradiction to $b \cdot y = \lambda d \cdot z$. Hence any such f must in fact send at least one of a, ..., z to zero.

On the other hand, from the way factors are paired off in (17) it is easy to verify that under any map into a ring without zero-divisors, if one of a, ..., z goes to zero, then either all of a, b, c, d, or all of x, y, z must go to zero, i.e., the map must be degenerate. ∎

Corollary 6.3. For any $\lambda \in K - \{0, 1\}$, $S(\lambda)$ is without zero-divisors, and in fact, remains without zero-divisors on tensoring with $\underline{U}(D)$, for any division algebra D.

Proof. Proper zero-divisors in $S(\lambda) \otimes \underline{U}(D)$ would correspond to a nondegenerate homomorphism of $S(\lambda)^* = S(\lambda^{-1})$ into $\underline{U}(D)$ (Lemma 3.1), which cannot exist by the above Lemma. ∎

So for any $\lambda \in K - \{0, 1\}$, $S(\lambda)$ and $S(\lambda)^* = S(\lambda^{-1})$ are bilinear systems without zero-divisors, whose tensor product has zero-divisors.

Yet the nature of this counterexample to question (4) strengthens our hope for (1) and (2), since the very proof that it was a counterexample showed that it could not show up inside a division algebra!

§7. Factoring functionals.

Since questions (3) and (4) have been answered negatively, if we hope to prove a positive answer to (1) or (2), we should look for special properties of some bilinear systems, that assure the nonexistence of zero-divisors in their tensor products.

In ring theory the most familiar "no zero divisors" result is the theorem that the polynomial ring R[X] over an integral domain R is again an integral domain. The proof is by looking at the leading terms $f_m X^m$ and $g_n X^n$ of two nonzero elements $f, g \in R[X]$, and noting that the X^{m+n} term of fg is $f_m g_n X^{m+n}$, which is nonzero because $f_m g_n \neq 0$ in R.

The same argument shows that group algebras of orderable groups have no zero-divisors; one can in fact generalize orderable to right orderable, and group to cancellative semigroup. Group theorists studying the (still open) question "Is a group algebra of a torsion-free group a ring without zero-divisors?" have still more generally defined a group G to have the u.p. (unique product) property if, given any two finite nonempty subsets $S, T \subseteq G$, there always exist $s \in S$, $t \in T$ such that the product st is "unique", i.e., $s't' \neq st$ for all $(s', t') \neq (s, t)$ in $S \times T$. If f, g are nonzero elements of a group algebra RG where R is without zero-divisors, and G has the u.p. property, let $S, T \subseteq G$ be their supports, and choose $s \in S$, $t \in T$ having unique product. Then the st term of fg is $f_s g_t st \neq 0$ and so RG is without zero-divisors [2], [3, §10]. (No example is known of a torsion-free group not having the u.p. property--or even the "2 u.p." property that whenever $|S \times T| \geq 2$, there are at least two unique products st and $\bar{s}\bar{t}$ in ST. If G has this stronger property and R is without zero-divisors, then all units in RG are monomials, ug, u a unit of R, $g \in G$.) Note in particular that if K is a field and G a group with the u.p. property, then KG remains without zero-divisors on tensoring over K with any K-algebra R without zero-divisors.

To get an analog of the above condition that is applicable to K-algebras or bilinear systems without distinguished bases, we should replace the conditions on finite subsets S and T by conditions on finite-dimensional subspaces. Note that the elements s, t and st used in stating the unique product property for groups were used to obtain, not elements, but <u>linear functionals</u> on the K-subspaces spanned by S, T and ST. The analog of the "unique product" property for groups will be the following "Factoring Functional" property for algebras or bilinear systems. Recall that K is an arbitrary field, and $\underline{C}_{inf}$ the category of non-necessarily finite-dimensional bilinear systems over K.

Lemma 7.1. For any $A \in Ob(\underline{C}_{inf})$, the following conditions are equivalent:

(a) (Factoring functional property). For all nonzero finite-dimensional subspaces $A_1' \subseteq A_1$, $A_2' \subseteq A_2$, there exist nonzero linear functionals α_1 on A_1', α_2 on A_2', α_3 on $\phi_A(A_1' \otimes A_2')$ such that for all $a \in A_1'$, $x \in A_2'$, $\alpha_3(\phi_A(a \otimes x)) = \alpha_1(a) \cdot \alpha_2(x)$.

(a_1') For all nonzero finite-dimensional subspaces $A_1' \subseteq A_1$, $A_2' \subseteq A_2$, there exists a subspace $A_1'' \subset A_1'$ of codimension 1 such that $\phi_A(A_1'' \otimes A_2') \neq \phi_A(A_1' \otimes A_2')$.

(a_2') Same, with roles of A_1' and A_2' reversed.

(a") Every finite-dimensional non-degenerate bilinear subsystem $A' \subseteq A$ has a nondegenerate map into the bilinear system $\underline{1}$.

Proof. The equivalence of (a) with each of the others is immediate. ∎

In fact, this condition is precisely what is needed to completely prevent zero-divisors in tensor products.

Proposition 7.2. For any $A \in Ob(\underline{C}_{inf})$, the following are equivalent:

(a) A has the factoring functional property (the equivalent conditions of Lemma 7.1).

(b) For every bilinear system B without zero-divisors, $A \otimes B$ is without zero-divisors.

If $A \in Ob(\underline{C})$ these are also equivalent to:

(c) Every non-degenerate homomorphic image of the dual system A^* has zero-divisors.

Proof. Assuming A finite-dimensional, we shall show (a) and (b) each equivalent to (c). For infinite-dimensional A, (a) and (b) each hold if and only if they hold for all finite-dimensional subsystems of A, hence will still be equivalent.

(a)<=>(c): Applying duality by "*", condition (a") of Lemma 7.1 translates to the condition that every non-degenerate precise homomorphic image of A* (i.e., dual of a subsystem of A--see last paragraph of §4) has a map of $\underline{1}$* into it, i.e., has zero-divisors. Now every homomorphic image of A* may be obtained from a precise homomorphic image by enlarging the 0-term and thus throwing additional relations into the 3-term, but leaving the 1- and 2-terms unchanged. This preserves the property of having zero-divisors. Hence (a") is indeed equivalent to (c).

(c)<=>(b): Recall that proper zero-divisors in $A \otimes B$ correspond to nondegenerate morphisms $A^* \to B$. Condition (c) says that no such morphisms can exist when B is without zero-divisors and is thus equivalent to (b). ■

Exercise 7.3. If A and B have the Factoring Functional property, so does $A \otimes B$. Prove this two ways, using condition (a) of Lemma 7.1, and using condition (b) of Proposition 7.2.

§8. Galois connections

The above is useful information but it does not show us how to get a handle on division algebras, in particular.

We have been asking what sort of conditions on bilinear systems imply that their tensor products have no zero-divisors. We shall now ask what sort of conditions are implied by embeddability in the bilinear system of a division algebra. The hope is that one can eventually link up conditions of these two sorts! Let us begin by recalling a ring-theoretic condition studied by A. A. Klein.

We need some notation and observations. If R is a ring and n a positive integer, let $R^n_{(\ell)}$ denote the free left R-module of rank n, represented as row vectors of length n with entries in R, and $R^n_{(r)}$ the free right R-module of rank n, represented by column vectors. Multiplication of row vectors by column vectors gives a map $R^n_{(\ell)} \times R^n_{(r)} \to R$, under which $R^n_{(\ell)}$ behaves as the module of all R-linear functionals on $R^n_{(r)}$, and vice versa. For every subset $X \subseteq R^n_{(\ell)}$, its annihilator Ann(X) is a submodule of $R^n_{(r)}$, and vice versa. Let $\underline{L}_n(R)$ denote the lattice of submodules of $R^n_{(\ell)}$ which are annihilators of subsets of $R^n_{(r)}$. (G.l.b.'s are given by intersections, and l.u.b.'s by double-annihilators of unions.) This is anti-isomorphic to the lattice of annihilator submodules in $R^n_{(r)}$. In formal terms, $\underline{L}_n(R)$ is the lattice of closed subsets under the Galois connection (cf. [4]) on $R^n_{(\ell)} \times R^n_{(r)}$ induced by the relation of annihilation.

Now I claim that if R is embeddable in a division ring, then for all $n \geq 0$ we have

(18_n) The lattice $\underline{L}_n(R)$ has length n.

I.e., all chains $0 = M_0 \subset M_1 \subset \ldots \subset M_m = R^n_{(\ell)}$ of annihilator submodules of $R^n_{(\ell)}$ satisfy $m \leq n$. This is clear if R _is_ a division ring, since the M_i are subspaces of the R-vector-space $R^n_{(\ell)}$. But note that whenever R S, the operations "closure in $S^n_{(\ell)}$" and "intersection with $R^n_{(\ell)}$", mapping $\underline{L}_n(R) \to \underline{L}_n(S) \to \underline{L}_n(R)$ are monotone and compose to the identity, so the length of $\underline{L}_n(R)$ is $\leq$ that of $\underline{L}_n(S)$. The assertion (18_n) when S is a division algebra follows. (Klein introduced (18_n) in the form "R is a domain, and every nilpotent $n \times n$ matrix over R has order of nilpotency $\leq n$." For the equivalence of these formulations see [5].) It was conjectured that (18_n)'s holding for all n was not only necessary but also sufficient for R to be embeddable in a division ring, but this was shown to be false in [6, Example 6.3, p. 51].

Now the most natural context for the above conditions is not rings, but bilinear systems! For any positive integer n, let $\underline{1}_n$ denote the bilinear system

$$0 \to \cdot \to \langle K^n, K^n \rangle \to K \to 0,$$

where the multiplication is "row vectors by column vectors". Given any A, let $\underline{L}(A)$ denote the lattice associated with the Galois connection on $A_1 \times A_2$ induced by the relation "$\phi_A(a \otimes x) = 0$".

<u>Definition 8.1.</u> Let A be a bilinear system and n a positive integer. We shall say that A satisfies $\underline{K}(n)$ ($\underline{K}$ for Klein) if the lattice $\underline{L}(A \otimes \underline{1}_n)$ has length $\leq n$. We shall say A satisfies $\underline{K}$ if it satisfies $\underline{K}(n)$ for all n.

We shall also want:

<u>Definition 8.2</u>. A bilinear system A will be called nonsingular if $\mathrm{Ann}(A_1)$ (a subspace of A_2) and $\mathrm{Ann}(A_2)$ (a subspace of A_1) are both zero; i.e., no nonzero element annihilates everything.

It is now easy to verify:

<u>Lemma 8.3</u>. A bilinear system embeddable in $\underline{U}(D)$ for some division algebra D over K satisfies condition $\underline{K}$, and is nonsingular.

A non-degenerate bilinear system A has no zero-divisors if and only if it is nonsingular and satisfies condition $\underline{K}(1)$. ∎

Let us show that the bilinear systems $S(\lambda)$ ($\lambda \neq 1$) constructed in §6, which were "bad" despite having no zero-divisors, fail to satisfy $\underline{K}$. The first $\underline{K}(n)$ which could fail for them is, by the above Lemma, $\underline{K}(2)$. And indeed, in A_1^2 we find the following length-3 chain of annihilators of subsets of A_2^2:

$$(19) \qquad \mathrm{Ann}(A_2^2) \subset \mathrm{Ann}(\{\binom{x}{y}, \binom{y}{z}\}) \subset \mathrm{Ann}(\{\binom{x}{y}\}) \subset \mathrm{Ann}(\emptyset).$$

To see that each of these inclusions is proper we note that the first annihilator is $\{0\}$, the second contains $(a, -c)$ and so is nonzero, an element in the third but not the second is $(b, -d)$, and the last term is all of A_1^2. We see that as in the proof of Lemma 6.2, it is the "skewed" relation between the first two columns of equations in (17) that leads to the impossibility of embedding $S(\lambda)$ in a division ring, in this case by causing the distinct second and third terms of (19). (The last column of (17) is used in Lemma 6.2 to exclude <u>non</u>-injective maps into division rings, but the existence of such maps is irrelevant to condition $\underline{K}$).

If we could somehow prove that for K algebraically closed, any two bilinear systems over K which satisfy K have tensor product satisfying K, we would have an affirmative answer to (2)! How the condition of algebraic closure could be used, however, is somewhat mysterious to me.

Conceivably, it might turn out that for K algebraically closed condition K implied the Factoring Functional property. This would be equivalent to an affirmative answer to the following question intermediate between (2) and (3):

($2\frac{1}{2}$) Let K be an algebraically closed field. If D is a division algebra over K, and R an associative K-algebra without zero-divisors, will $D \otimes_K R$ be without zero-divisors?

On the other hand, the Factoring Functional condition does not imply K. To get a counterexample, delete the last defining relation from (17). The resulting system A has $m = 4$, $n = 3$, $p = 7$. But we saw in §5 that for these values of m and n, $p = 6$ was the only case in which a bilinear system without zero-divisors over an algebraically closed field could fail to have property (b) of Proposition 7.2. Hence this system must satisfy that condition, which is equivalent to the Factoring Functional property. Nonetheless, (19) shows that A does not satisfy K(2).

We remark that adjoining an additional relation to the 3-component of a bilinear system (going to a monoepimorphic image) in general neither preserves nor reflects condition K; though it reflects the Factoring Functional property.

Condition K is concerned with tensoring with the bilinear systems $\underline{1}_n$, so the following observations relating such systems to arbitrary systems might prove useful. Let $\underline{1}_{(m,n,p)}$ denote the bilinear system based on multiplication of $m \times n$ matrices and $n \times p$ matrices to get

$m \times p$ matrices. For any positive m, n, p and bilinear system A, it is easy to check that $\underline{L}(A \otimes \underline{1}_{(m,n,p)}) \cong \underline{L}(A \otimes \underline{1}_n)$. (Key point: two matrices have product zero if and only if every row of the first annihilates every column of the second.) On the other hand, I claim that any nonsingular bilinear system B can be embedded in a system $\underline{1}_{(p,n,1)}$. Namely, take $n = \dim B_2$, $p = \dim B_3$, and represent elements of B_2 and B_3 as col. vectors. Elements of B_1 then induce, via ϕ_A, linear maps from B_2 to B_3, and so can be represented by $p \times n$ matrices.

§9. A problem of J. Lewin.

The following question was raised by J. Lewin ("with no rhyme or reason". Personal communication.) He asked it for rings, but that form reduces to the same question for algebras over prime fields, so one might as well pose it for algebras over arbitrary fields:

(20) Let K be a field and r a positive integer. Does there exist an associative K-algebra R such that $R^{\otimes r}$ (= $R \otimes \ldots \otimes R$, r times) does not have zero-divisors, but $R^{\otimes(r+1)}$ does?

By the methods used in §2, the above is easily shown equivalent to:

(21) Let K be a field and r a positive integer. Does there exist a finite-dimensional bilinear system A over K such that $A^{\otimes r}$ does not have zero-divisors, but $A^{\otimes(r+1)}$ does?

If $r = 1$ we know the answer is affirmative. I suspect that it is for all r. A further reduction of this problem will be made at the end of §12.

In the next four sections, 10-13, we shall mostly forget our troublesome zero-divisors, and examine for its own sake some of the rich structure of the category of bilinear systems over a field K.

The last six sections, 14-19, comprise an appendix dealing with a number of topics related to our zero-divisors problem which I did not include above to avoid interrupting the main thread. We shall see in §14 that (1)-(3) have affirmative answers if either of the algebras is commutative.

All the remaining sections are essentially independent of one another except that sections 15 and 17 require 14.

MORE ABOUT $\underline{C}$

§10. $\otimes$ and $\otimes^*$ ....

We saw in Section 3 that for any finite-dimensional bilinear system A, the dual system A^* is characterized by the existence of a universal pair of zero-divisors in $A \otimes A^*$; equivalently, a universal member of $\mathrm{Hom}(\underline{1}^*, A \otimes A^*)$. The universality means that for all systems B:

(22) $\mathrm{Hom}\ (A^*, B) \cong \mathrm{Hom}\ (\underline{1}^*, A \otimes B)$.

If in (22) we put A^* in place of A, we get

(23) $\mathrm{Hom}\ (A, B) \cong \mathrm{Hom}\ (\underline{1}^*, A^* \otimes B)$.

This means the bifunctor $(A, B) \to A^* \otimes B$ is an "internal hom" for the category $\underline{C}$ with respect to the set-valued functor $\mathrm{Hom}\ (\underline{1}^*, -)$.

Now starting with three bilinear systems A, B, C one has $\mathrm{Hom}(B, A \otimes C) \cong \mathrm{Hom}(\underline{1}^*, B^* \otimes (A \otimes C)) \cong \mathrm{Hom}(\underline{1}^*, (B^* \otimes A) \otimes C) \cong \mathrm{Hom}((B^* \otimes A)^*, C) \cong \mathrm{Hom}(B \otimes^* A^*, C) \cong \mathrm{Hom}(A^* \otimes^* B, C)$. In summary

(24) $\mathrm{Hom}(B, A \otimes C) \cong \mathrm{Hom}(A^* \otimes^* B, C)$,

which says the functor $A \otimes -$ has the left adjoint $A^* \otimes^* -$. Put another way, if A is a bilinear system, we can not only find one system with a universal map of $\underline{1}^*$ into its tensor product with A (to wit, A^*), but for any B, we can find a system with a universal map of B into its tensor product with A (namely $A^* \otimes^* B$).

Returning to our internal hom--how is composition of homomorphisms represented? Given bilinear systems A, B and C and elements $f \in \mathrm{Hom}(\underline{1}^*, A^* \otimes B)$, $g \in \mathrm{Hom}(\underline{1}^*, B^* \otimes C)$, these induce by functoriality a morphism $f \otimes^* g$ from $\underline{1}^* = \underline{1}^* \otimes^* \underline{1}^*$ to

$$(A^* \otimes B) \otimes^* (B^* \otimes C). \tag{25}$$

Now it is not hard to show that this object has a map into

$$A^* \otimes (B \otimes^* B^*) \otimes C \tag{26}$$

which acts as the identity on the 1- and 2-parts (cf. next section). There is also a canonical map $B \otimes^* B^* \to \underline{1}$ (acting as trace on the 1- and 2-parts, dual to the universal map $\underline{1}^* \to B^* \otimes B$), so a map is induced from (26) to $A^* \otimes \underline{1} \otimes C \cong A^* \otimes C$. Composing these functorial maps with $f \otimes^* g$, we get a map $\underline{1}^* \to A^* \otimes C$, corresponding to $g \circ f : A \to C$.

§11. ... and their friends.

If A is a bilinear system, and U, V are finite-dimensional vector spaces, we can form a bilinear system

$$0 \to U \otimes A_0 \otimes V \to \langle U \otimes A_1, A_2 \otimes V\rangle \to U \otimes A_3 \otimes V \to 0.$$

which we will call $A_{(U,V)}$.

If B is another bilinear system, let us abbreviate $A_{(B_1, B_2)}$ to A_B. This is a very trivial construction with many trivial properties. It is clearly functorial in A and B. Also:

$$A_C \otimes B \cong (A \otimes B)_C \cong A \otimes B_C, \text{ and } A_C \otimes^* B \cong (A \otimes^* B)_C \cong A \otimes^* B_C \tag{27}$$

$$A_B \otimes C \cong A_{B \otimes^* C} \cong (A_B)_C \cong A_{B_C}. \tag{28}$$

We shall denote the common value of (28) by A_{BC}. From (27) we also see that

$$(29)\qquad \underline{1}_B \otimes A \cong A_B \cong \underline{1}^*_B \otimes^* A$$

(where $\underline{1}^*_B$ means $(\underline{1}^*)_B$, not $(\underline{1}_B)^*$). Finally, note that $\underline{1}^*_A \cong \underline{1}^* \otimes A$. Hence by (29), A_B can be written $A \otimes^* (\underline{1}^* \otimes B)$. Symmetrically, $\underline{1}_A \cong \underline{1} \otimes^* A$, and so $A_B \cong A \otimes (\underline{1} \otimes^* B)$.

Note that the functors $(U,V) \mapsto \underline{1}_{(U,V)}$ and $(U,V) \mapsto \underline{1}^*_{(U,V)}$ from $(K\text{-}\underline{\text{vector-space}})^2$ to $\underline{C}_{inf}$ are respectively the left and the right adjoint to the "forgetful" functor $A \mapsto (A_1, A_2)$. The counit and unit respectively of these adjunctions are the natural maps shown on the left-hand side of (30) below. The 1- and 2-components of both these maps are the identity maps of A_1 and A_2.

Bilinear systems with 1- and 2-components A_1 and A_2:	Their respective 0-components as subspaces of $A_{12} =_{def.} A_1 \otimes A_2$:

$$(30)\qquad \begin{array}{c} \underline{1}^*_A \\ \uparrow \\ A \\ \uparrow \\ \underline{1}_A \end{array} \qquad\qquad \begin{array}{c} A_{12} \\ | \\ A_0 \\ | \\ \{0\} \end{array}$$

If A is a bilinear system, let us (as on p. 10, and also (30) above) abbreviate $A_1 \otimes A_2$ to A_{12}, and think of A as determined by A_1, A_2, and a subspace $A_0 \subseteq A_{12}$. We saw in §4 that for two systems A and B the systems $A \otimes B$ and $A \otimes^* B$ both have the same 1-component $A_1 \otimes B_1$ and 2-component $A_2 \otimes B_2$, but that their 0-components are distinct subspaces of $A_{12} \otimes B_{12}$, namely $A_{12} \otimes B_0 + A_0 \otimes B_{12}$ and $A_0 \otimes B_0$ respectively. We now note that the systems A_B, B_A, $\underline{1}_{AB}$ and $\underline{1}^*_{AB}$ also have these same 1- and 2-components. The 0-components of all six systems, and the natural maps among these systems determined by the inclusions among the 0-components, are shown in (31):

(31)

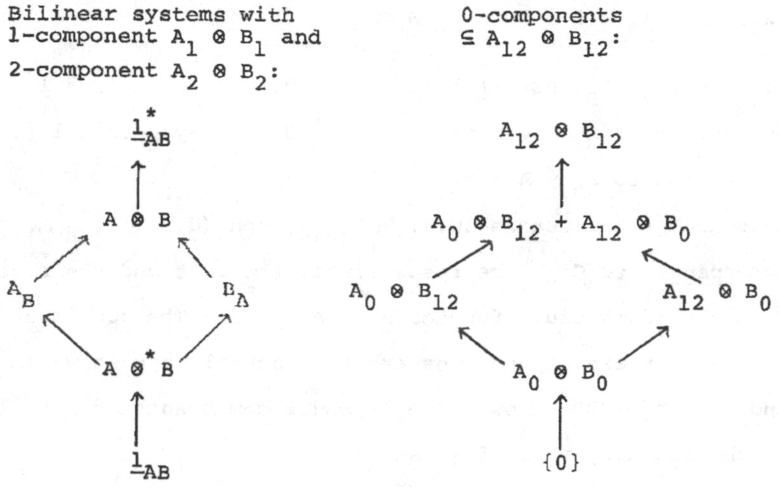

Note that the right-hand side of (31) is a <u>lattice</u> of subspaces of $A_{12} \otimes B_{12}$, under sum and intersection. Hence the left-hand diagram is "cartesian", in the sense that the diagram-theoretic least upper bound W of any two objects X and Y will be their pushout over any lower bound Z, and likewise greatest lower bounds are pullbacks. So, for instance, $A \otimes B$ can be written as the pushout of

$\underline{1}_{AB} \begin{smallmatrix} \nearrow A_B \\ \searrow B_A \end{smallmatrix}$, and $A \otimes^* B$ as the pullback of $\begin{smallmatrix} A_B \searrow \\ B_A \nearrow \end{smallmatrix} \underline{1}^*_{AB}$. Thus, our

"nontrivial" constructions $\otimes$ and $\otimes^*$ can be obtained in this sense from the "trivial" constructions $\underline{1}$, $\underline{1}^*$ and A_B.

If we start with n bilinear systems, A, B, ..., E, the diagram analogous to (31) takes the form of the <u>free distributive lattice with 0 and 1 on</u> n generators. These generators are represented by $A_{BC\ldots E}$, $B_{AC\ldots E}$, ..., $E_{AB\ldots D}$; or in terms of 0-components, $A_0 \otimes B_{12} \otimes \ldots \otimes E_{12}$, $A_{12} \otimes B_0 \otimes \ldots \otimes E_{12}$, ..., $A_{12} \otimes B_{12} \otimes \ldots \otimes E_0$. (The key step in verifying this description of the lattice generated by the above n subspaces is to take bases for A_{12}, B_{12}, etc., which respectively extend bases of A_0, B_0, etc., and use the induced basis of $A_{12} \otimes \ldots \otimes E_{12}$, noting that each of the subspaces mentioned is spanned by a subset of the latter basis.) The case n = 3 is shown in (32):

(32)

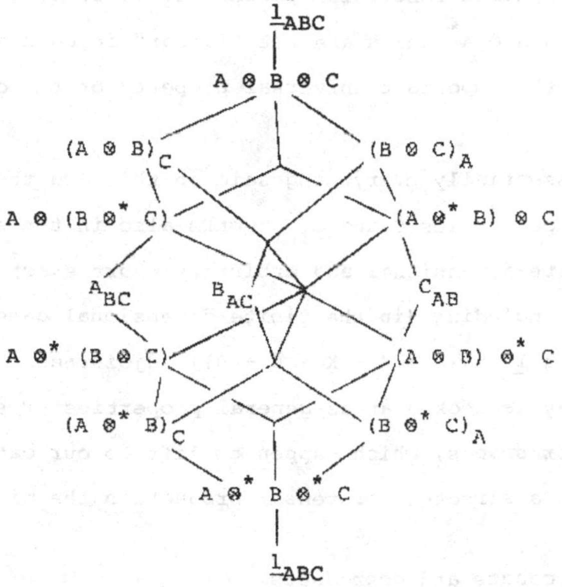

Some of the vertices are left unlabeled for legibility's sake. These can be easily identified using symmetry, except for the middle vertex of the diagram, which represents a new construction, not factorizable into $\otimes$ etc. It has 0-component $A_0 \otimes B_0 \otimes C_{12} + A_0 \otimes B_{12} \otimes C_0 + A_{12} \otimes B_0 \otimes C_0$. (Compare with $A \otimes B \otimes C$, which has $A_0 \otimes B_{12} \otimes C_{12} + A_{12} \otimes B_0 \otimes C_{12} + A_{12} \otimes B_{12} \otimes C_0$, and with $A \otimes^* B \otimes^* C$, which has $A_0 \otimes B_0 \otimes C_0$.)

We can read off from (32) such facts as that $A \otimes (B \otimes^* C)$ is not in general isomorphic to $(A \otimes B) \otimes^* C$, but that there is a natural map from the latter to the former. Among the maps appearing in the corresponding diagram for n=4 is a map $(A \otimes B) \otimes^* (C \otimes D) \to A \otimes (B \otimes^* C) \otimes D$

which we used in getting from (25) to (26) in describing composition of "internal homs" in $\underset{\sim}{C}$! (The inequality in the free distributive lattice on 4 generators to which it corresponds is $(w \vee x) \wedge (y \vee z) \leq w \vee (x \wedge y) \vee z$. Cf. the logical proposition $((P \vee Q) \wedge (R \vee S)) \Rightarrow (P \vee (Q \wedge R) \vee S)$.)

I have not explored functorial properties of constructions such as $(A \otimes B) \otimes^* A$ and $A \otimes A^*$ which are not "linear" in each argument, except for noting the important universal property of the canonical map $\underline{1}^* \to A \otimes A^*$.

Actually, essentially everything said in this and the preceding section about the categories $\underset{\sim}{C}$ and $\underset{\sim}{C}_{\mathrm{inf}}$ works also in the simpler categories of finite-dimensional and arbitrary short exact sequences of vector-spaces, including (in the finite-dimensional case) the internal hom (using $\underline{1}^* = (0 \to K \to K \to 0 \to 0)$) adjointness, etc. So these phenomena may be looked at as general properties of short exact sequences of vector spaces, which happen to lift to our categories, i.e., which respect a structure of tensor product on the middle term.

§12. Matrices, products and coproducts.

Suppose A and B are two bilinear systems and that $A_1 = B_1 = V$. Then let us define (A, B) to be the system $0 \to A_0 \oplus B_0 \to \langle V, A_2 \oplus B_2 \rangle \to A_3 \oplus B_3 \to 0$. Likewise, if A and C are systems such that $A_2 = C_2$, we can combine them into a system $\binom{A}{C}$ whose 2-component is the common value of A_2 and C_2. These two modes of combination can be joined, so that, for instance, we get a system

$$\begin{pmatrix} A & B \\ C & D \end{pmatrix} =_{\mathrm{def.}} \begin{pmatrix} (A\ B) \\ (C\ D) \end{pmatrix} = \left(\binom{A}{C} \binom{B}{D}\right)$$

whenever we have four bilinear systems such that the 1-components of the pairs of systems we have written as row coincide, and the 2-components of the systems we have written as columns. Larger $m \times n$ "matrices" of bilinear systems are, of course, defined the same way. We note that $\begin{pmatrix} A & B \\ C & D \end{pmatrix}$ will have zero-divisors if and only if at least one of the component systems does.

If A and B are two bilinear systems, we can now easily describe their coproduct and product in $\underline{C}$. The coproduct will be

$$\begin{pmatrix} A & \underline{1}(A_1, B_2) \\ \underline{1}(B_1, A_2) & B \end{pmatrix}$$

or explicitly,

$$0 \to \begin{pmatrix} A_0 & 0 \\ 0 & B_0 \end{pmatrix} \to \langle \begin{pmatrix} A_1 \\ B_1 \end{pmatrix}, (A_2, B_2) \rangle \to \begin{pmatrix} A_3 & A_1 \otimes B_2 \\ B_1 \otimes A_2 & B_3 \end{pmatrix} \to 0$$

and the product will be the same with $\underline{1}^*$ in place of $\underline{1}$.

One can see from the above description of the coproduct that the set of pairs of proper zero-divisors in the coproduct of A and B can be naturally identified with the disjoint union of the corresponding sets for A and for B!

With the help of the above constructions, we can make another reduction of Lewin's problem of §9.

Lemma 12.1. Questions (20) and (21) have an affirmative answer (for given K and r) if and only if there exist bilinear systems $A^1, \ldots, A^{r+1}$ over K, such that the (r+1)-fold tensor product $A^1 \otimes \ldots \otimes A^{r+1}$ has zero-divisors, but none of the r-fold products $A^{i_1} \otimes \ldots \otimes A^{i_r}$ $(i_1, \ldots, i_r \leq r+1$, not necessarily distinct) have zero-divisors.

Sketch of Proof. If A satisfies (21), then we can take all A^i to be A.

On the other hand, if $A^1, \ldots, A^{r+1}$ exist having the above properties, we take for A their coproduct in $\underline{C}$. The above description of coproducts, plus the obvious distributive laws $(A,B) \otimes C \cong (A \otimes C, B \otimes C)$ etc., plus our observation on when such matrices of systems have zero-divisors, show that this A will satisfy the condition of (21). ■

The conditions on $A^1, \ldots, A^{r+1}$ in the above "reduction" seem more complicated than the original hypotheses on A in (21)--for instance they require that <u>each</u> A^i should, like the one A of (21), have $(A^i)^{\otimes r}$ without zero-divisors. The simplification lies in the subtle point that no bilinear system is now required to have zero-divisors in a tensor product with copies of itself. In particular, the following approach to the reduced problem looks hopeful. Choose any "reasonable" systems for $A^1, \ldots, A^r$, and then take $A^{r+1} = (A^1 \otimes\ldots\otimes A^r)^*$. This forces zero-divisors in the (r+1)-fold tensor product, but there is no apparent reason for them to appear in any tensor product of fewer terms. A variant approach: Choose a sequence of bilinear systems $B^1, \ldots, B^r$; also define $B^0 = \underline{1}$, $B^{r+1} = \underline{1}^*$. Let $A^i = B^{i-1*} \otimes^* B^i$. Each A^i has by construction the property (cf. comments following (24)) that its tensor product with B^{i-1} contains a canonical copy of B^i. It follows by induction that $A^1 \otimes\ldots\otimes A^i$ has a copy of B^i, and in particular that $A^1 \otimes\ldots\otimes A^{r+1}$ contains a copy of $\underline{1}^*$, i.e., a pair of zero-divisors. Again, there is no reason for smaller tensor products to have zero-divisors. What one must do to make either of these ideas work is to show that one can choose the initial objects A^i or B^i so that the zero-divisors that "have no reason to occur" really don't! Possibly a "general position" method like that of §5 could be used.

(Remark: The systems $S(\lambda)$ constructed in §6 cannot be used as any of the A^i in this construction if $r > 1$ and κ is algebraically closed. For we showed that $S(\lambda)^* \cong S(\lambda^{-1})$; but taking $\lambda = \kappa^3$, one also gets $S(\lambda^{-1}) \cong S(\lambda)$ via $a \mapsto b$, $b \mapsto a$, $c \mapsto \kappa d$, $d \mapsto \kappa c$, $x \mapsto x$, $y \mapsto \kappa^{-1} y$, $z \mapsto \kappa z$. Hence $S(\lambda) \otimes S(\lambda)$ has zero-divisors, so a family of bilinear systems involving an $S(\lambda)$ cannot satisfy the hypotheses of Lemma 12.1 for $r > 1$.)

§13. Alternative viewpoints and variant categories.

There are a number of rather different ways of looking at the objects of our categories $\underline{C}$ and C_{inf}. It is worthwhile to note some of these, because as we shall see, one viewpoint can suggest constructions and approaches which are not as evident from another. Above we mainly used three closely related viewpoints: that of a bilinear map $A_1 \times A_2 \to A_3$; that of a short exact sequence with a tensor product for middle term (which was our formal definition); and that of a pair of vector spaces A_1, A_2 with a distinguished subspace of tensors, $A_0 \subseteq A_1 \otimes A_2$.

Let us make explicit a different point of view which we used briefly at the end of §8: a <u>nonsingular</u> bilinear system A is determined up to natural isomorphism by the two vector-spaces A_2 and A_3, and the vector-space of linear maps $\bar{A}_1 \subseteq \mathrm{Hom}(A_2, A_3)$, obtained from A_1 by defining $\bar{a}(x) = \phi_A(a, x)$ for each $a \in A_1$, $x \in A_2$. The necessary and sufficient conditions for three spaces $A_2 \xrightarrow{\bar{A}_1} A_3$ to correspond to such a bilinear system is that the intersection of the kernels of the elements of $\bar{A}_1$ be zero, and the sum of their images be A_3; hence we may think of nonsingular bilinear systems as systems of two spaces and a space of maps with this property. Note that duality of finite-dimensional vector-spaces by Hom(-, K) will carry a 3-tuple of this sort to another such 3-tuple, yielding a "duality" construction on bilinear systems distinct from our functor $(\ )^*$! Another construction this viewpoint suggests is the following. Let A and B be systems such that $A_3 = B_2$; thus we have vector spaces and spaces of maps $A_2 \xrightarrow{\bar{A}_1} A_3 = B_2 \xrightarrow{\bar{B}_1} B_3$. Let $\bar{B}_1\bar{A}_1 \subseteq \mathrm{Hom}(A_2, B_3)$ denote the space spanned by compositions of the maps in the above families; a homomorphic image of $\bar{B}_1 \otimes \bar{A}_1$. These will comprise the 1-component of a new bilinear system, the "composition" of A and B, which in our exact sequence form looks like $0 \to \cdot \to \langle B_1A_1, A_2\rangle \to B_3 \to 0$. Still another bilinear system implicit in this situation is $0 \to \cdot \to \langle B_1, A_1\rangle \to B_1A_1 \to 0$.

If A is a nonsingular finite-dimensional bilinear system, note that the space $\mathrm{Hom}(A_2, A_3)$ can be identified with $A_2^* \otimes A_3$. Hence our space-of-linear-maps description of A leads to the "subspace of a tensor-product" description of a new system

$$0 \to A_1 \to \langle A_2^*, A_3 \rangle \to \cdot \to 0.$$

What is this construction good for? I don't know.

Concerning the "$A_0 \subseteq A_1 \otimes A_2$" idea of a bilinear system--these are in effect the objects considered in [13], where I studied the number $r(A) = \inf_{u \in A_0 - \{0\}} rk(u)$ (rk = rank as a tensor); this can also be described as the least positive integer r such that there exists a monomorphism $\underline{1}_r^* \to A$. (Lemma 2.1 above can be generalized to say that the K-algebra K<A> will have n-term weak algorithm if and only if n is less than r(A).)

Another area in which we might keep an open mind is the way we have made our bilinear systems into a category. The definition of $\underline{C}$ was natural and convenient, but there are alternatives which might be advantageous for one or another kind of consideration.

For example, we might identify two morphisms $f, f': A \to B$ of $\underline{C}$ whenever there exist nonzero scalars $\alpha_i \in K (i = 0, 1, 2, 3)$ such that $f_i' = \alpha_i f$. Hom(A,B) - {0} then becomes a complete algebraic variety. Or we might only identify f and f' if there exists $\alpha \in K - \{0\}$ such that $f_0' = f_0$, $f_1' = \alpha f_1$, $f_2' = \alpha^{-1} f_2$, $f_3' = f_3$. Then the set of <u>non-degenerate</u> morphisms $A \to B$ can be identified with the set of nonzero linear maps $A_1 \otimes A_2 \to B_1 \otimes B_2$ that carry reducible tensors to reducible tensors and A_0 into B_0. Or we might define Hom(A, B) to be precisely the set of all linear maps with the above property; this differs from the preceding definition in that all degenerate morphisms become identified.

In another direction, we could drop the exactness condition at one or both ends of (8). In particular, if in the definition of $\underline{C}_{inf}$ we drop exactness on the right (thus returning to our original concept of bilinear system from §1), the category of such systems is a variety of algebras in the sense of universal algebra [4]. (Except that these will be 3-sorted algebras, i.e., will have a 3-tuple of underlying sets A_1, A_2, A_3. Some universal algebraists allow such objects, others do not. However we can get an isomorphic variety of 1-sorted algebras by using $A_1 \oplus A_2 \oplus A_3$ as underlying set, and introducing the projection maps among our operations.) Thus the methods of universal algebra become available if we use this category.

It is not clear what category structures if any may be natural to our $A_2 \xrightarrow{\bar{A}_1} A_3$ concept of a bilinear system. We would like the "duality" we noted for such systems to become a functor, but this is difficult as it is contravariant in A_2 and A_3 but covariant in $\bar{A}_1$. Some possibilities are (1) a category with "relations" ($f_i \subseteq A_i \times B_i$) for morphisms; (2) a category which treats A_2 and A_3 differently, e.g., with morphisms;

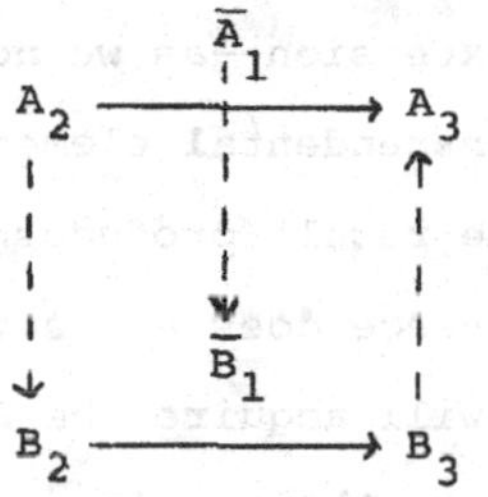

(3) to use several category structures, and let "duality" permute these.

We now return to our zero-divisors problem, with some miscellaneous results and ideas.

APPENDICES

14. Commutative K-algebras.

It is known that questions (1) - (3) have affirmative answers for <u>commutative</u> algebras. In fact we can show

<u>Lemma 14.1</u>. Let K be an algebraically closed field and C a commutative K-algebra without zero-divisors. Then $\underline{U}(C)$ has the Factoring Functional property.

Proof. We can clearly reduce to the case where C is finitely generated as a K-algebra. By algebraic geometry [7, p. 335, Prop. 3] the variety Spec C has a nonsingular point $\underline{p}$; i.e., there is a maximal ideal $\underline{p}$ of C such that the graded ring $gr_{\underline{p}}(C)$ associated with the $\underline{p}$-adic filtration of C is a polynomial ring over K [7, p. 329, Definition 2, and discussion following]. This filtration on C is separating (Krull's Theorem). From the fact that polynomial algebras satisfy the Factoring Functional property, it is easy to deduce that C does. ∎

Thus, for K and C as above, K-algebras and K-bilinear systems without zero-divisors remain without zero-divisors on extending scalars to C.

If K is not algebraically closed, then $\underline{U}(C)$ need not have the Factoring Functional property, even if C is a field extension of K containing no proper algebraic extension--as we noted in §1, if we adjoin to the real numbers a transcendental element X, and an element Y satisfying $X^2 + Y^2 + 1 = 0$, the result produces zero-divisors on tensoring with the quaternions, hence does not satisfy the Factoring Function property. However, it will acquire the Factoring Functional property on extension of scalars to the complexes! In general we have:

Corollary 14.2. Let K be a perfect field, with algebraic closure $\overline{K}$, and let C be a field extension of K in which K is algebraically closed. Then $\underline{U}(C \otimes \overline{K})$ has the Factoring Functional property as a bilinear system over $\overline{K}$.

Proof. By the preceding result this will follow if $C \otimes \overline{K}$ is without zero-divisors. Clearly it will suffice to show $C \otimes L$ without zero-divisors for all finite algebraic extensions L of K, and since K is perfect we can apply the Theorem of the Primitive Element and write $L = K(\alpha)$. By Lemma 1.1 we need only prove that the minimal polynomial P of α over K remains irreducible over C. But any proper factors of P over C must have some coefficients not in K and these would be properly algebraic over K, contradicting the algebraic closure of K in C. ∎

The above Corollary can fail for non-perfect K: If $K(a^{1/p}, b^{1/p})$ is non-primitive, then $C = K(x, (ax^p + b)^{1/p})$ is a counterexample.

§15. Two comparison lemmas.

The following result shows that the kind of structure carried by a bilinear system is nearly as "strong" as that carried by a general associative K-algebra.

<u>Lemma 15.1</u>. Let K be a field, R an associative K-algebra, and D a division algebra over K. Then every non-degenerate homomorphism of bilinear systems, $f:\underline{U}(R) \to \underline{U}(D)$ arises as follows: take elements $u, v \in D - \{0\}$, and an algebra homomorphism $h:R \to D$, and set $f_1(a) = u\,h(a)$, $f_2(x) = h(x)\,v$, $f_3(p) = u\,h(p)\,v$.

<u>Proof</u>. Given f, let $u = f_1(1)$, $v = f_2(1)$. Then $u \neq 0$, else we would have $0 = u\,f_2(R) = f_1(1)\,f_2(R) = f_3(R) = f_1(R)\,f_2(R)$ in D, so $f_1(R) = 0$ or $f_2(R) = 0$, contradicting our assumption of non-degeneracy. Likewise, $v \neq 0$. So write $f_1' = u^{-1}f_1$, $f_2' = f_2\,v^{-1}$, $f_3' = u^{-1}f_3v^{-1}$. We see that f' is also a homomorphism $\underline{U}(R) \to \underline{U}(D)$, and $f_1'(1) = f_2'(1) = 1$. Now for any $a \in R$, $f_1'(a) = f_1'(a)f_2'(1) = f_3'(a) = f_1'(1)f_2'(a) = f_2'(a)$. So $f_1' = f_2' = f_3'$, and the equations stating that f' is a homomorphism of bilinear systems now show that the common value of f_1', f_2', f_3' is a K-algebra homomorphism h. ∎

The following result is also pleasant:

<u>Lemma 15.2</u>. Let K be a field, R a K-algebra which is a right Ore ring, and D its skew field of fractions. Then $\underline{U}(D)$ has the Factoring Functional property if and only if $\underline{U}(R)$ does.

Proof. Assume $\underline{U}(R)$ has the property, and let A be a finite-dimensional subsystem of $\underline{U}(D)$. We can take a common right denominator for the elements of A_1, and so write $A_1 = B_1 r^{-1}$ with $B_1 \subseteq R$, and then a common right denominator for $r^{-1}A_2$, so that $r^{-1}A_2 = B_2 s^{-1}$. Then we find that the bilinear system A is isomorphic to a subsystem $B \subseteq \underline{U}(R)$ via the map $a \mapsto ar$ $(a \in A_1)$, $x \mapsto r^{-1}xs$ $(s \in A_2)$, $p \mapsto ps$ $(p \in A_3)$. The desired functional can be found for B, and hence also for A. The converse implication is trivial. ∎

§16. On Procesi's original question.

Let me point out one approach applicable specifically to question (1) on division algebras finite-dimensional over their centers.

Let D and E be two such division algebras, with centers C(D) and C(E). From Lemma 14.1 we know that $C(D) \otimes C(E)$, $D \otimes C(E)$, and $C(D) \otimes E$ will all be K-algebras without zero-divisors. In particular, $C(D) \otimes C(E)$ is a commutative domain; call it C and let K' denote its field of fractions. The other two tensor-product rings will be central C-algebras, free of finite ranks as C-modules, hence they will be Ore rings and their fields of fractions can be seen to be $D' = D \otimes_{C(D)} K'$ and $E' = K' \otimes_{C(E)} E$, finite-dimensional central division algebras over K'.

Now the ring $D \otimes_K E$ will be a C-order in the central simple K'-algebra $D' \otimes_{K'} E'$; hence it will be without zero-divisors if and only if the latter is a division algebra, a question in Brauer group theory.

Notice that if we choose a basis for D over C(D), and a basis for E over C(E), and write down structure constants for D and E as algebras over their centers in terms of these bases, then these become structure constants for D', respectively E', as algebras over K'. Hence D' and E' have structure constants in subfields of K' which are linearly disjoint over K. One feels that this should somehow prevent the kind of "interaction" between the factors that leads to zero-divisors!

One way one might use this linear disjointness is to show that the subfields C(D) and C(E) of K' may be mapped essentially independently by homomorphisms or specializations of fields over K, and consider the effects of such maps on the Brauer group of K', and in particular on the elements representing D' and E'.

I don't know whether every field between K' and D' has a K'-basis with respect to which its structure constants lie in C(D)--probably not.

The following reduction has been pointed out to me by several people:

<u>Lemma 16.1</u>. Question (1) is equivalent to the special case where [D:C(D)] and [E:C(E)] are powers of a common prime.

<u>Proof</u>. By [8, Theorem 4.10], D and E can each be written as a tensor product over its center of division algebras D_{p_i}, E_{p_i} of prime-power degrees. The special case referred to would assure that all $D_{p_i} \otimes_K E_{p_i}$ are without zero-divisors, i.e., that the $D'_{p_i} \otimes_{K'} E'_{p_i}$ are division algebras. Since these have relatively prime degrees, their tensor product, $D' \otimes_{K'} E'$, will be a division algebra by [8, Theorem 5.10]. ∎

§17. Zero-divisors => nilpotents.

Adam Hausknecht has shown me the following result:

<u>Proposition 17.1 (Hausknecht)</u>. Let D and E be arbitrary division algebras over an algebraically closed field K. If $D \otimes_K E$ has no nilpotents, then it has no zero-divisors. ∎

His proof can be adapted to give a more general statement. To formulate this we recall that a prime ring R is called <u>centrally closed</u> if for all nonzero a, b ε R such that

$\forall\, r \in R,\ arb = bra,$

there exists $\alpha \in C(R)$ such that $a = \alpha b$. If R is centrally closed, then $C(R)$ is a field. Any prime ring R has a central closure R', a centrally closed prime over-ring with the properties that $R' = R\,C(R')$, and for all $\alpha \in C(R')$, there exist $a \in R$, $b \in R - \{0\}$ with $a = \alpha b$. $C(R')$ is called the <u>extended center</u> of R, which we shall write $C'(R)$. (See Martindale [9]. He speaks of the "extended centroid", but for rings with 1, centroid = center, so I am using the latter term.)

The following result implies Proposition 17.1 above because a prime ring with zero-divisors has nilpotents.

<u>Proposition 17.2</u>. Let R and S be prime algebras over a field K, with extended centers $C'(R)$, $C'(S)$. Then $R \otimes_K S$ is prime if and only if $C'(R) \otimes_K C'(S)$ has no zero-divisors. In particular, if K is algebraically closed, then $R \otimes_K S$ <u>is</u> prime.

<u>Proof</u>. First, suppose $C'(R) \otimes_K C'(S)$ does have proper zero-divisors, $(\sum \alpha_i \otimes \beta_i)(\sum \gamma_j \otimes \delta_j) = 0$. One can choose nonzero $w, y \in R$, $x, z \in S$ such that all $\alpha_i w$ and $\gamma_j y$ lie in R, and all $\beta_i x$ and $\delta_j z$ lie in S. Then $\sum \alpha_i w \otimes \beta_i x$ and $\sum \gamma_j y \otimes \delta_j z$ are proper zero-divisors in $R \otimes S \subseteq R' \otimes S'$.

On the other hand, suppose that $C'(R) \otimes_K C'(S)$ has no zero-divisors. We shall show that $R' \otimes_K S'$ is prime; this implies that $R \otimes_K S$ is prime, since the former ring is generated over the latter by central elements.

The key fact we need first is that in a tensor product $U \otimes_K V$ of centrally closed prime K-algebras, every nonzero two-sided ideal I contains a nonzero element of the form $a\, u \otimes v$, where $a \in C(U) \otimes_K C(V)$, $u \in U$, and $v \in V$. To see this, choose a nonzero element $p = \sum_m u_i \otimes v_i \in I$, whose rank m is minimal. Note that the families (u_i) and (v_i) in U and V are each linearly independent over K. Now for every $z \in U$, the element $u_1 z p - p z u_1 = \sum (u_1 z u_i - u_i z u_1) \otimes v_i$ belongs to I, and has rank < m (the i = 1 term is zero), hence by choice of m, is zero. This means that for all $z \in U$ and all i,

$u_1 z u_i - u_i z u_1 = 0$. As U is centrally closed we can write each u_i as $\alpha_i u_1$ ($\alpha_i \in C(U)$). Likewise, we can write $v_i = \beta_i v_1$. So $p = (\sum \alpha_i \otimes \beta_i) u_1 \otimes v_1$, which is of the desired form. (Cf. Herstein [12, Lemma 4.1.1, p. 90], from which Hausknecht got this idea, and Martindale [9, Theorem 2.2].)

Now, returning to $R' \otimes S'$, with $C(R') \otimes C(S')$ an integral domain, suppose I and J are nonzero two-sided ideals of $R' \otimes S'$. Let us choose, by the above, nonzero elements
$a\, w \otimes x \in I$, $b\, y \otimes z \in J$ ($a, b \in C(R') \otimes C(S')$, $w, y \in R'$, $x, z \in S'$). Since R and S are prime we can find $r \in R$, $s \in S$ such that wry and xsz are nonzero. I claim that the element of IJ:

$$(a\, w \otimes x)(r \otimes s)(b\, y \otimes z) = ab\ (wry \otimes xsz)$$

is nonzero. Indeed, if we choose a $C(R')$-basis for R' containing the nonzero element wry, and a $C(S')$-basis for S' containing xsz, these will induce a $C(R') \otimes C(S')$-basis for $R' \otimes_K S'$ containing $wry \otimes xsz$. Since this appears with nonzero coefficient ab in the above expression, that expression is nonzero, $IJ \neq \{0\}$, so $R' \otimes S'$ is prime. ∎

§18. The case m = 2.

We found in §5 that for K algebraically closed, no zero-divisors could occur in tensor-products $A \otimes B$ of bilinear systems without zero-divisors unless

(16) $\qquad (m-2)(n-2) \geq 2 \qquad (m = \dim A_1,\ n = \dim A_2)$.

In terms we used later, this means that any A without zero-divisors such that $(m-2)(n-2) < 2$ must have the Factoring Functional property. In particular, this will hold when m = 2, n arbitrary. I shall give here a direct proof of this, since it throws a little light on the still-mysterious question of how the assumption "K algebraically closed" can be used in studying general bilinear systems. The result we shall prove is in fact a little better than what we already know:

Lemma 18.1. Let K be *any* field, and A a bilinear system over K, with $\dim A_1 \leq 2$, which does not satisfy the Factoring Functional condition. Then after extension of scalars to the algebraic closure of K, A has zero-divisors.

Proof. Clearly we may assume A is non-degenerate, is finite-dimensional, and is minimal for the hypothesis, i.e., that every proper subsystem of A has the Factoring Functional property. Thus the conditions of Lemma 7.1 must fail with $A' = A$.

In particular, condition (a_1') fails, i.e.:

(33) There is no subspace $A_1'' \subseteq A_1$ of codimension 1, such that $\phi_A(A_1'' \otimes A_2) \neq A_3$.

Clearly (33) excludes the possibility $\dim A_1 = 1$, so let $\{a,b\}$ be a basis of A_1, and let $\bar{a}$, $\bar{b}$ denote the induced maps $A_2 \to A_3$ (see §13). Then (33), applied to the 1-dimensional subspace of A_1 spanned by a says that $\bar{a}$ is surjective. If $\bar{a}$ is not also injective then a is a zero-divisor in A and we are done. If it is injective, it is an isomorphism $\bar{a}$: $A_2 \cong A_3$. In particular $\det(t\bar{a} - \bar{b})$ (computed with respect to arbitrary bases of A_2 and A_3) is a polynomial of positive degree, hence has a root λ in the algebraic closure of K. In the induced bilinear system over this algebraic closure, $\lambda a - b$ will be a zero-divisor. ■

(I might more simply have described λ as an eigenvalue of $b\,a^{-1} \in \operatorname{End} A_3$, but I thought the above formulation might extend better to higher-dimensional cases, where one might, say, want to look at the variety $\{(\alpha, \beta, \gamma) | \det(\alpha a + \beta b + \gamma c) = 0\}$. Incidentally, the conclusion of the above Lemma is false as stated for $\dim A_1 = \dim A_2 = 3$, even though (16) still fails. For a counterexample, let C be the commutative $\mathbb{R}$-algebra mentioned just before Corollary 14.2, and take $A \subseteq \underline{U}(C)$ with

$A_1 = A_2 =$ space spanned by $\{1, X, Y\}$.)

§19. Condition $\underline{K}$, and the embedding problem for semigroups.

Recall that our proof that a bilinear system $S(\lambda)$ $(\lambda \neq 1)$ could not be mapped non-degenerately into the multiplicative system of a division algebra involved, as a key step, calculating that it could not be embedded (in a weak sense not involving K-linearity) in the multiplicative semigroup of a group. Later essentially the same calculation came up in the verification that this system did not satisfy $\underline{K}$. This suggests that condition $\underline{K}$, or some analog thereof with sets in place of vector spaces, might be relevant to the question of when a semigroup is embeddable in a group. A very elegant set of necessary and sufficient condition for such embeddability is in fact already known, due to A. I. Mal'cev [10], [4, Theorem 7.3.3]. But that is no reason not to look at other conditions.

Given a ring R, the condition $\underline{K}(n)$ says that every chain of annihilator-submodules of the left R-module $R^n_{(\ell)}$--submodules determined by families of linear relations $a_1x_1 +\ldots+ a_nx_n = 0$ on the coordinates a_i--should have length $\leq n$. If S is a semigroup, we can analogously form the left S-set $S^n_{(\ell)}$, but we obviously cannot write down the above sort of relations. The kind of right-linear relation that we can speak of for such n-tuples is

$$(34) \qquad a_i x = a_j x' \qquad (i, j \leq n;\ x, x' \in S).$$

Hence let us define the right S-set $S^{<n>}_{(r)} = \{(i,j;\ x,x') \mid i,j \leq n;\ x,x' \in S\}$ and the relation

$Z_{(\ell)} = \{((a_1, \ldots, a_n), (i, j; x, x')) \mid a_i x = a_j x'\} \subseteq S^n_{(\ell)} \times S^{<n>}_{(r)}$.

This relation induces a Galois connection between $S^n_{(\ell)}$ and $S^{<n>}_{(r)}$. Let $\underline{L}_n(S)_{(\ell)}$ denote the lattice of subsets of $S^n_{(\ell)}$ closed with respect to this connection--solution-sets of families of equations (34).

Let G be a group. We now need an analog of dimension theory for members of $\underline{L}_n(G)_{(\ell)}$.

Given a finite family $(a_i)_{i\varepsilon I}$ of elements of G, let us define their <u>right ratio</u> to be the I^2-tuple of elements $\rho_{ij} = a_i^{-1}a_j$ (the elements by which one must right-multiply the various terms of this family to get the others.) This data is very redundant--an appropriate $|I|-1$ of these elements determine all the rest via the formulae

(35) $\rho_{hj} = \rho_{hi}\rho_{ij}$,

but we will keep this redundant data for the sake of symmetry. Note that the right-ratio of a family (a_i) is unchanged by left translation of this family; in fact, it constitutes a full set of invariants for such families under left translation. Note also that given a family of elements $(a_i)_{I \cup I'}$, the right ratios of the subfamilies $(a_i)_I$ and $(a_i)_{I'}$ determine the right ratio of the whole family <u>if</u> $I \cap I' \neq \emptyset$.

<u>Lemma 19.1</u>. Let G be a nontrivial group, and $U \varepsilon \underline{L}_n(G)_{(\ell)}$. Then either

(i) $U = \emptyset$, or

(ii) there exists a partition of $\{1, \ldots, n\}$ into disjoint families $I_1, \ldots, I_r$, and for each I_m a family of elements of G, $(\rho_{ij})_{i,j\varepsilon I_m}$ satisfying (35), such that

$$U = \{(a_1, \ldots, a_n) \mid (\forall m \leq r)(\forall i, j \varepsilon I_m)\ a_i^{-1}a_j = \rho_{ij}\}.$$

Let us write dim U = 0 in the first case, and dim U = r in the second. Then dim U is well-defined, and $U \subsetneqq V \varepsilon \underline{L}_n(G)_{(\ell)}$ implies dim U < dim V.

Sketch of Proof. Clearly $U = G^n_{(\ell)}$ itself satisfies (ii), with $r = n$.

Now suppose U is a set satisfying (ii), and we impose on it a new relation (34), i.e., take $(i, j; x, x') \in G^{<n>}_{(r)}$ and let

$V = \{(a_i) \in U \mid a_i x = a_j x'\}$. Then there are three possibilities: first, the indices i and j may lie in different class I_m and $I_{m'}$ under the partition involved in the definition of U. In this case it is straightforward that for $(a_i) \in V$, the right ratio of $(a_i)_{i \in I_m \cup I_{m'}}$ is uniquely determined, and that in fact V will have the form (ii) with the same partition as for U, except that I_m and $I_{m'}$ become united into one class. On the other hand, if i and j lie in the same class I_m, then the right ratio implied by (34), $a_i^{-1} a_j = x\,x'^{-1}$ may either agree with the ratio ρ_{ij} involved in the description of U--in which case V = U--or may be different, in which case our conditions are inconsistent, and $V = \emptyset$.

Note that a set U satisfying (ii) uniquely determines the partition $I_1, \ldots, I_r$. Namely, i_1 and i_2 belong to the same subset if and only if $\forall (a_i), (b_i) \in U$, $a_{i_1} = b_{i_1} \Rightarrow a_{i_2} = b_{i_2}$.

The remaining details of the proof are straightforward. ∎

As for rings, we deduce:

Lemma 19.2. If n is a positive integer, and S is a nontrivial semigroup embeddable in a group, then the lattice $\underline{L}_n(S)_{(\ell)}$ has length n. ∎

We shall say a semigroup S satisfies $K(n)_{(\ell)}$ if $\underline{L}_n(S)$ has length n.

Since this definition is not left-right symmetric, one also has the obvious dual condition, $\underline{K}(n)_{(r)}$. We shall write $\underline{K}_{(\ell)}$ and $\underline{K}_{(r)}$ for the conditions $\forall n\ \underline{K}(n)_{(\ell)}$ and $\forall n\ \underline{K}(n)_{(r)}$.

It is easy to show that $\underline{K}(1)_{(\ell)}$ is equivalent to right cancellation. $\underline{K}(2)_{(\ell)}$ is what was involved in our argument about $S(\lambda)$; in fact, one finds that $\underline{K}(2)_{(\ell)}$ (and also, by symmetry, $\underline{K}(2)_{(r)}$) fails in any semigroup S with elements a, b, c, d, w, x, y, z satisfying

$$ax = cy \qquad aw = cz$$
$$bx = dy \qquad bw \neq dz.$$

(In particular, taking $a = c = w = z = 1$, $x = y$, we see that $\underline{K}(2)_{(\ell)} \Rightarrow$ left cancellation $= \underline{K}(1)_{(r)}$.)

<u>Lemma 19.3</u>. Let n be a positive integer and S a nontrivial semigroup. Then the following conditions are equivalent:

(i) S satisfies $\underline{K}(n)_{(\ell)}$.

(ii) For each 2n-tuple of elements of S, $(x_1, x_2', x_2, x_3', \ldots, x_n, x_1')$ one has either

$$(\forall\, a_1, \ldots, a_n \in S)(a_1x_1 = a_2x_2' \wedge \ldots \wedge a_{n-1}x_{n-1} = a_nx_n') \Rightarrow (a_nx_n = a_1x_1')$$

<u>or</u>

$$(\forall\, a_1, \ldots, a_n \in S)(a_1x_1 = a_2x_2' \wedge \ldots \wedge a_{n-1}x_{n-1} = a_nx_n') \Rightarrow (a_nx_n \neq a_1x_1')$$

<u>Sketch of Proof</u>. (i) => (ii) is straightforward, by looking at the chain of subsets of $S^n_{(\ell)}$ defined by successively larger subfamilies of our n equations, and also the subset $\emptyset$. Conversely, assume (ii). Note that from this condition we can get the corresponding conditions for all $m \leq n$, by taking $x_i = x_{i+1}'$ for $n - m$ values of i. Now consider the set $U \subseteq S^n_{(\ell)}$ defined by some family X of relations of the form (34). Draw a graph whose set of vertices is $\{1, \ldots, n\}$, with an edge connecting any two vertices i and j which occur as the indices in one of the relations in X. Suppose now that we adjoin one more relation to X. Then it follows from (ii) that <u>if</u> the associated indices i and j lie in the same component of the graph, the new relation is either <u>implied</u> by some family of the old ones (any chain of length $m \leq n$ connecting i and j), or it is <u>inconsistent</u> with such a family. Hence whenever we add a relation to X, we either decrease the number of components in our graph, or we do not affect our solution-set, or the

latter becomes empty. This easily yields condition $\underline{K}(n)_{(\ell)}$; cf. proof of Lemma 19.1. ■

The systems of equations occurring in Lemma 19.2 (ii) can be conveniently diagrammed. Given $x_1, \ldots, x_n; x_1', \ldots, x_n'$, draw a 2n-gon, and label its edges as follows:

(36)

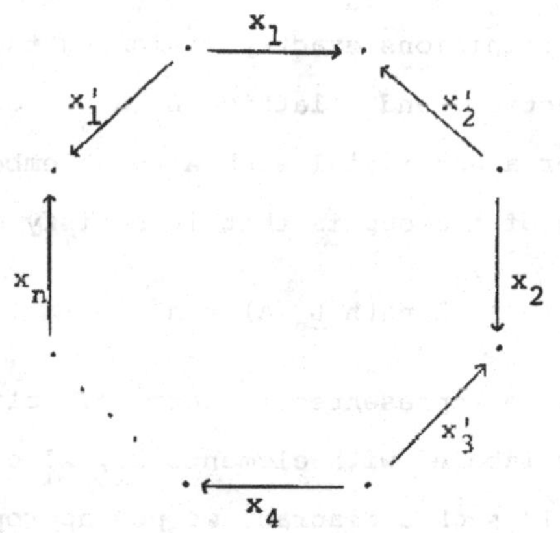

We now try to label each vertex of this diagram with an element of S, in such a manner that if we have an edge labeled $\overset{a}{\cdot} \xrightarrow{x} \overset{b}{\cdot}$, then $ax = b$. Condition (ii) above says that if a diagram (36) can have its vertices labeled in at least one consistent way, then any other labelling that is satisfactory at all but one edge will be satisfactory at all edges.

This means that in a semigroup S satisfying $\underline{K}_{(\ell)}$, all diagrams (36) can be divided into three disjoint classes: (a) those for which a consistent labelling of the vertices exists; (b) those for which a labelling exists which is satisfactory at 2n-1 of the edges, but unsatisfactory at the remaining edge; and (c) those for which there is no labelling satisfactory at 2n-1 edges. It is easy to see that if S is embedded in a group G, then for any diagram (36), we have $x_1x_2'^{-1}x_2 \ldots x_nx_1'^{-1} = 1$ if (a) holds, $x_1x_2'^{-1} \ldots x_1'^{-1} \neq 1$ if (b) holds, while either may be possible if (c) holds. The existence of the last case was to be expected because we know there are semigroups S which can be embedded as generating subsemigroups in more than one non-isomorphic overgroups. Let us call the diagram (36) <u>fulfillable</u> if it satisfies condition (a), and <u>antifulfillable</u> in case (b).

We stop here to sketch how to generalize these considerations to structures analogous to our bilinear systems. Define a <u>binary system</u> to be a 4-tuple $A = (A_1, A_2, A_3; \phi_A)$, where A_i are <u>sets</u>, and ϕ_A is a surjective set-map $A_1 \times A_2 \to A_3$. For each n, ϕ_A induces a relation $Z_{(\ell)} \subseteq A_1^n \times A_2^{<n>}$ (definitions exactly mimicking the semigroup case) and hence a Galois connection, and a lattice $L_n(A)_{(\ell)}$ of subsets of A_1^n. A necessary condition for a nontrivial such A to be embeddable in the underlying binary system of a group is that it satisfy

$$\underline{K}(n)_{(\ell)}: \qquad \text{length } \underline{L}_n(A) = n$$

This condition can again be represented in terms of polygonal diagrams (36); here the edges are labeled with elements x_i, $x_i' \in A_2$, and to "fulfill" or "antifulfill" such a diagram, we put appropriate elements of A_1 at "source" vertices, and of A_3 at "sink" vertices. It is easy to show that a semigroup S is embeddable in a group G if and only if its underlying binary system $\underline{U}(S)$ is embeddable in $\underline{U}(G)$ (cf. Lemma 12.1), and clearly S satisfies $\underline{K}_{(\ell)}$ if and only if $\underline{U}(S)$ does.

To investigate the sufficiency of $\underline{K}_{(\ell)}$ for embeddability of a binary system in a group, let A be a binary system and let us consider the set of formal expressions $uv^{-1}\ldots yz^{-1}$ (u,...,z an even number of elements in A_2). Let us write $uv^{-1}\ldots yz^{-1} \sim pq^{-1}\ldots st^{-1}$ ($p,\ldots,t \in A_2$, possibly a different even number of terms) if the diagram with edges labeled u, v,..., y, z, t, s,..., q,p is fulfillable. We then ask: (a) is ~ an equivalence relation on these formal expressions? (b) does it respect formal multiplication of expressions? and (c) if one or both of the above questions has a negative answer, and we let ≈ denote the least multiplication-respecting equivalence relation on such expressions which contains ~, then will at least the following condition, obviously necessary for the embeddability of A in a group hold?

(37) $\forall\, u, \ldots, z \in A_2$, $uv^{-1}\ldots yz^{-1} \approx 1$ implies that the diagram (36) associated with $uv^{-1}\ldots yz^{-1}$ is not <u>antifulfillable</u>.

Unfortunately, the answer to all these questions is "no". I shall sketch the reason, leaving the detailed arguments to the interested reader. The key point is that we can construct a binary system A satisfying $\underline{K}_{(\ell)}$, and having elements u, v, w, x, y, z $\in A_2$ such that the perimeters of the top and bottom cells of (38) below are fulfillable, but the perimeter of the whole diagram is antifulfillable.

(38)

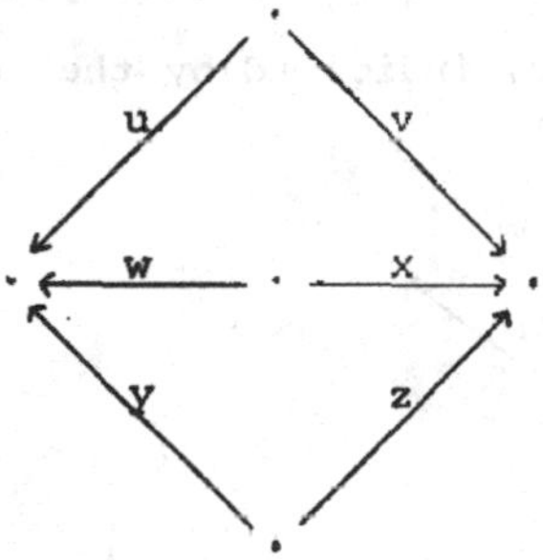

Namely, we take $A_1 = \{a, b, c, d, e, f\}$, $A_2 = \{u, v, w, x, y, z\}$, and A_3 to be the quotient of the set $A_1 \times A_2$ by the relations

(39)
$$\begin{array}{lll} au = bw & cw = dy & ey = fu \\ av = bx & cx = dz & \end{array}$$

Note that we do <u>not</u> set

(40) $ez = fv$.

To see that A satisfies $K_{(\ell)}$, one notes that the only fulfillable cycles in A are

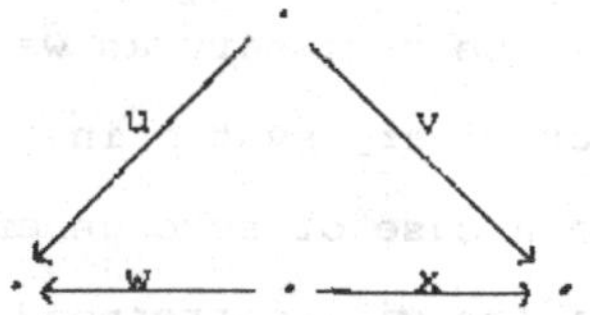

and

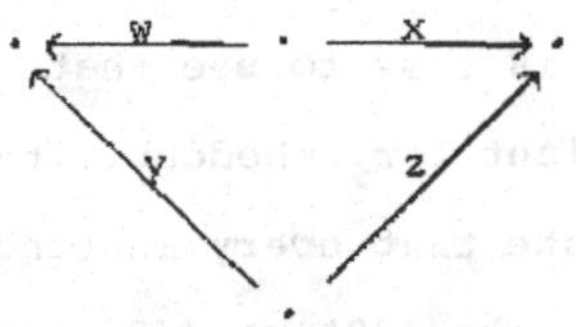

(and cycles obtained from these by "backtracking"), and neither of the above two cycles is also antifulfillable.

Thus, a condition which must be satisfied by any binary system embeddable in a group, but which is not implied by $\underline{K}_{(\ell)}$, is $(\forall a, \ldots, f \in A_1, u, \ldots, z \in A_2)$ (39) => (40). (More generally, any five of the equations (39) $\cup$ (40) implies the sixth. I believe that the universal enveloping semigroup <A> of the above binary system A will likewise be an example of a semigroup satisfying $\underline{K}$ but not embeddable in a group.)

It happens that the above example fails to satisfy the dual condition $\underline{K}_{(r)}$. This is shown by the following cycle, and the two ways of labelling its source-vertices, indicated by the letters on the inside and outside respectively:

(41)

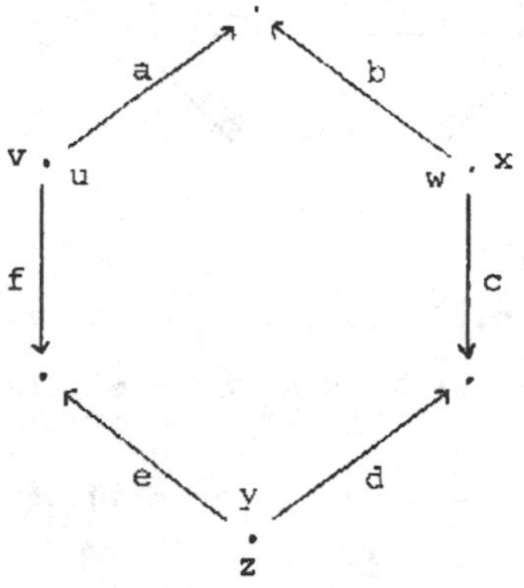

This, however, is an accident. If one modifies (38) by adding another pair of links to any one of the "bridges", one gets an example satisfying both $\underline{K}_{(\ell)}$ and $\underline{K}_{(r)}$ but not (37). (The enlarged version of (38) also has a "dual" diagram, like (41), with edges in A_1 and alternate vertices in A_2 and A_3. But this is no longer a simple cycle.)

It is easy to see that condition (37) will be necessary as well as sufficient for embeddability of a semigroup or binary system in a group. Note that every instance of (37) will be a case of some general condition like (39) => (40), and each such condition will correspond to a diagram like (38). It would be interesting to investigate whether there is a natural way to translate between such diagrams and Mal'cev's conditions ([10], [4]).

It is not hard to deduce from the unsolvability of the word problem for groups that there can be no algorithm for determining (in finitely many steps) whether a given finite binary system is embeddable in a group!

Note that if A is a binary system and K a field, then we can form a bilinear system KA over K, letting $(KA)_i$ (i = 1, 2, 3) be a K-vector-space with basis A_i, and ϕ_{KA} be the K-linear extension of ϕ_A. One immediately sees that

(41) (KA satisfies condition $\underline{K}$) => (A satisfies $\underline{K}_{(\ell)}$ and $\underline{K}_{(r)}$).

The reverse implication is false, e.g., if A = $\underline{U}$(G), where G is a group with torsion elements. It would be interesting to study the left-hand side of (41) as a condition on A and K.

I originally thought that the considerations of this section might lead to new conditions on bilinear systems embeddable in division rings, which could help us in attacking our original problems (1) and (2). However, because a relation (34) in a semigroup or binary system can involve only two terms, the nonlinear version of $\underline{K}$ which we obtained above was much weaker than the linear version. So it is not clear that our strengthenings of our nonlinear condition have analogs which are nontrivial strengthenings of our original version. However, perhaps these ideas could be of some interest to semigroup theory.

REFERENCES

1. R. Gordon (ed.), Ring Theory (Proceedings of a conference on ring theory held in Park City, Utah, March 2-6, 1971), Academic Press, 1972.

2. B. Banaschewski, "On proving the absence of zero-divisors for semi-group rings," Canad. Math. Bull. 4(1961) 225-231.

3. D. S. Passman, Advances in group rings, Israel J. Math. 19(1974) 67-107.

4. P. M. Cohn, Universal Algebra, Harper & Row (1965).

5. A. A. Klein, "A remark concerning embeddability of rings in fields," J. Alg. 21(1972) 271-274.

6. G. M. Bergman, "Coproducts and some universal ring construction," Trans. Amer. Math. Soc. 200(1974) 33-88.

7. D. Mumford, Introduction to Algebraic Geometry, preliminary version of first three chapters, Harvard (1966).

8. A. A. Albert, Structure of Algebras, Amer. Math. Soc. Colloquium Publications, No. XXIV (1939).

9. W. S. Martindale, 3d, "Prime rings with involution and generalized polynomial identities," J. Alg. 22(1972) 502-516.

10. A. I. Mal'cev, "On the embedding of associative systems in groups," I (Russian, German summary) Mat. Sbornik 6(48)(1939) 331-336.

11. P. M. Cohn, Universal skew fields of fractions, Symposia Mathematica 8(1972) 135-148.

12. I. H. Herstein, Noncommutative Rings, Carus Mathematical Monographs, No. 15, Mathematical Association of America, 1968.

13. G. M. Bergman, Ranks of tensors and change of base-field," J. Alg. 11(1969) 613-621.

REGULAR RINGS AND RANK FUNCTIONS

K. R. Goodearl
University of Utah
Salt Lake City, Utah

This paper is concerned with the existence and uniqueness of rank functions (in the sense of von Neumann) on a von Neumann regular ring R, and with the structure of the completion of R with respect to the metric induced by a rank function. Section 1 discusses several situations in which the existence (and/or uniqueness) of rank functions is known. Section 2 studies a generalization (pseudo-rank functions), with the help of which some fairly general existence theorems can be proved. Section 3 discusses the structure of completions, and Section 4 lists a number of open questions related to the ideas discussed in this paper.

Since the majority of the proofs in this paper have been written up in detail elsewhere, the proofs presented here are either sketched or omitted.

All rings in this paper are associative with unit, and ring maps are assumed to preserve the unit.

§1. Rank Functions

<u>Definition</u>. A (<u>normalized</u>) <u>rank</u> <u>function</u> [17, p. 231] on a regular ring R is a map $N:R \to [0,1]$ such that

(a) $N(x) = 0$ if and only if $x = 0$.

(b) $N(1) = 1$.

(c) $N(xy) \leq N(x), N(y)$.

(d) $N(e + f) = N(e) + N(f)$ for orthogonal idempotents e, f.

An obvious consequence of (c) is that if $x, y \in R$ with $xR = yR$, then $N(x) = N(y)$. More generally, if $xR \cong yR$, then $N(x) = N(y)$. [Since

R is assumed to be regular, $xR \cong yR$ implies that $x = ayb$ and $y = cxd$ for suitable $a, b, c, d \in R$, in which case it follows from (c) that $N(x) \leq N(y)$ and $N(y) \leq N(x)$.] A less obvious consequence of (c) and (d) is that $N(x + y) \leq N(x) + N(y)$ for all $x, y \in R$ [17, p. 231].

The most basic example of a rank function, and the source of the definition, comes from the usual definition of rank for matrices. Thus let F be a field, and let R be the ring of all $n \times n$ matrices over F, for some positive integer n. Conditions (a), (c), (d) of the definition above are satisfied by matrix rank, hence we may define a rank function N on R by normalizing: $N(x) = \text{rank}(x)/n$ for all $x \in R$. It is not hard to check that this is in fact the only rank function on R.

More generally, there is a rank function on any simple Artinian ring R. Namely, set $N(x) = \ell(xR)/\ell(R)$ for all $x \in R$, where $\ell(-)$ denotes length (in terms of composition series). Then N is a rank function on R, and is the only one [4, Proposition 2].

Rank functions (under a different name) also occur in measure theory. For if R is a Boolean algebra (made into a ring using symmetric difference and infimum for sum and product), then a rank function on R is exactly a positive, finitely additive, probability measure on R.

Lemma 1.1 [4, Lemma 1]. Let R be a regular ring such that $R = R_1 \times \ldots \times R_t$ with each $R_i \neq 0$. If we have a rank function N_i on R_i for each i, together with positive real numbers $\alpha_1, \ldots, \alpha_t$ such that $\alpha_1 + \ldots + \alpha_t = 1$, then we can define a rank function N on R according to the rule

$$N(x_1, \ldots, x_t) = \alpha_1 N_1(x_1) + \ldots + \alpha_t N_t(x_t).$$

Conversely, every rank function on R is of this form.

Proof. It is trivial to check that any such N is a rank function.

Conversely, let N be any rank function on R. If e_i denotes the unit of R_i, then $e_1, \ldots, e_t$ are orthogonal idempotents in R such that $e_1 + \ldots + e_t = 1$. As a result, we obtain real numbers $\alpha_i = N(e_i)$ such that $\alpha_1 + \ldots + \alpha_t = 1$.

We view R_i as a (non-unital) subring of R, and set $N_i(x_i) = N(x_i)/\alpha_i$ for all $x_i \in R_i$. Then N_i is a rank function on R_i, and N is derived from the α_i and the N_i as stated in the hypotheses. ■

The construction of rank functions given in Lemma 1.1 can also be carried out for countably infinite products. Thus let $R_1, R_2, \ldots$ be nonzero regular rings, and set $R = \Pi R_i$. If we have a rank function N_i on R_i for each i, and positive real numbers $\alpha_1, \alpha_2, \ldots$ such that $\sum \alpha_i = 1$, then we can define a rank function N on R according to the rule $N(x_1, x_2, \ldots) = \sum \alpha_i N_i(x_i)$. We note one property of this N: if e_i denotes the unit of R_i, then $\sum N(e_i) = \sum \alpha_i = 1$.

Conversely, however, there may exist rank functions on a countably infinite direct product which cannot be obtained in this manner. For example, let $F_1, F_2, \ldots$ be fields, and set $R = \Pi F_i$. There is a unique rank function N_i on F_i: namely, N_i is the characteristic function of the set $F_i - \{0\}$. Choose a maximal ideal M of R which contains $\oplus F_i$, and define a map $P: R \to \{0, 1\}$ to be the characteristic function of the set R - M. Then the rule

$$N(x_1, x_2, \ldots) = P(x_1, x_2, \ldots)/2 + \sum N_i(x_i)/2^{i+1}$$

defines a rank function N on R. If e_i denotes the unit of F_i, then $\sum N(e_i) = 1/2$, hence N cannot be obtained as in the preceding paragraph.

Combining Lemma 1.1 with our observations above on the unique rank function on a simple Artinian ring, we can completely describe the rank functions on a semisimple Artinian ring, as follows.

Proposition 1.2 [4, Proposition 2]. Let $R = R_1 \times \ldots \times R_t$, where each

R_i is a simple Artinian ring. If $\alpha_1, \ldots, \alpha_t$ are positive real numbers such that $\alpha_1 + \ldots + \alpha_t = 1$, then we can define a rank function N on R according to the rule

$$N(x_1, \ldots, x_t) = \alpha_1 \ell(x_1R_1)/\ell(R_1) + \ldots + \alpha_t \ell(x_tR_t)/\ell(R_t).$$

Conversely, every rank function on R is of this form.

If R is a regular ring with a rank function, then Lemma 1.1 shows that any nonzero ring direct summand of R must have a rank function. However, a nonzero factor ring of R need not have a rank function. For example, let $F_1, F_2, \ldots$ be fields, and set $R = \Pi F_i$. As shown above, there exist rank functions on R. However, $R/(\oplus F_i)$ contains uncountable direct sums of nonzero ideals, from which it follows that there cannot exist a rank function on $R/(\oplus F_i)$. Another example is constructed in [10, Example 1], which shows that if R is a regular ring with a rank function, prime factor rings of R need not have rank functions.

By contrast, rank functions do carry over from a regular ring R to any full matrix ring over R. In order to discuss this, we must relate rank functions to the finitely generated projective R-modules.

Definition. For any ring R, let FP(R) denote the full subcategory of Mod-R generated by all finitely generated projective right R-modules.

Proposition 1.3. If R is a regular ring, then FP(R) is an abelian category in which every morphism splits.

Proof. It is clear that FP(R) has finite products and coproducts. According to [12, Lemma 4], any finitely generated submodule of any $A \in FP(R)$ is a direct summand of A, from which it follows that every morphism in FP(R) splits, and that FP(R) has kernels and cokernels. Also, any subobject in FP(R) is the kernel of a projection onto a direct summand, and likewise any quotient object is a cokernel. ■

Definition. A module A is subisomorphic to a module B, written $A \lesssim B$, provided A is isomorphic to a submodule of B.

Definition. Let R be a regular ring. A (normalized) dimension function on FP(R) is a map d from FP(R) into the nonnegative real numbers such that

(a) $d(A) = 0$ if and only if $A = 0$.

(b) $d(R) = 1$.

(c) If $A \lesssim B$, then $d(A) \leq d(B)$.

(d) $d(A \oplus B) = d(A) + d(B)$.

For example, if R is a field, then vector space dimension defines a dimension function on FP(R).

Proposition 1.4 [10, Proposition 4]. Let R be a regular ring. If d is any dimension function on FP(R), then the rule $N_d(x) = d(xR)$ defines a rank function N_d on R. Conversely, if N is any rank function on R, there exists a unique dimension function d on FP(R) such that $N_d = N$.

Proof. It is clear that $N_d(x) = 0$ if and only if $x = 0$, that $N_d(1) = 1$, and that $N_d(e + f) = N_d(e) + N_d(f)$ for orthogonal idempotents e, f. Given x, y ε R, we have $xyR \leq xR$, whence $N_d(xy) \leq N_d(x)$. Also, $xyR \lesssim yR$ (because xyR is projective), whence $N_d(xy) \leq N_d(y)$. Thus N_d is a rank function on R.

Now consider any rank function N on R. Given A ε FP(R), we have $A \cong x_1R \oplus \ldots \oplus x_nR$ for suitable x_i ε R [12, Theorem 4], and we set $d(A) = N(x_1) + \ldots + N(x_n)$. It must be checked that d is well-defined, and that it is a dimension function on FP(R). It is clear that $N_d = N$, and that d is unique. ■

Corollary 1.5 [8, Theorem 1]. Let R and S be Morita-equivalent regular rings. Then there is a rank function on R if and only if there is a rank function on S.

Proof. Since FP(R) is equivalent to FP(S), this follows directly from Proposition 1.4. ■

Definition. A module A is directly finite provided A is not isomorphic to any proper direct summand of itself. It is easily checked that A is directly finite if and only if all one-sided inverses in End(A) are two-sided, i.e., fg = 1 implies gf = 1. In particular, R_R is directly finite if and only if xy = 1 implies yx = 1 in R, if and only if ${}_RR$ is directly finite. In this case we say that R is a directly finite ring. Note that any direct limit of directly finite rings must be directly finite. [This is not true for modules, of course. For example, an infinite-dimensional vector space (which is not directly finite) is the direct limit of its finite-dimensional subspaces (which are directly finite).]

Note that any direct summand of a directly finite module is directly finite.

A ring Morita-equivalent to a directly finite ring need not be directly finite, as [15, Theorem 1.0] shows. For regular rings, the question of the Morita-invariance of direct finiteness is open.

Definition. We shall use $M_n(R)$ to denote the ring of all n × n matrices over a ring R.

Lemma 1.6. If R is a regular ring with a rank function, then $M_n(R)$ is directly finite for all n.

Proof. By Corollary 1.5, there is a rank function on each $M_n(R)$, hence it suffices to show that R is directly finite. If xy = 1 in R, then yx and 1 - yx are orthogonal idempotents such that yx + (1 - yx) = 1. There is a rank function N on R, and N(yx) + N(1 - yx) = 1. Observing that x(yx)y = 1, we see that N(yx) = 1. Thus N(1 - yx) = 0, whence yx = 1. ■

<u>Definition</u>. A regular ring R is said to satisfy the <u>comparability axiom</u> [6] provided that for all x, y ε R, either $xR \lesssim yR$ or $yR \lesssim xR$. Equivalently, R satisfies the comparability axiom if and only if for all x, y ε R, there exist a, b ε R such that either $x = ayb$ or $y = axb$. This latter formulation shows that the comparability axiom is left-right symmetric, and that it is preserved under direct limits. For example, [3, Lemma 5', p. 832] shows that any prime, regular, right self-injective ring satisfies the comparability axiom. On the other hand, [4, Example A] shows that not every simple regular ring satisfies the comparability axiom.

Proposition 1.7 [6, Lemma 3.7]. Let R be a regular ring which satisfies the comparability axiom. For any A, B ε FP(R), either $A \lesssim B$ or $B \lesssim A$.

Corollary 1.8. Let R and S be Morita-equivalent regular rings. Then R satisfies the comparability axiom if and only if S satisfies the comparability axiom.

<u>Definition</u>. For any module A and any nonnegative integer k, we use kA (rather than A^k) to denote the direct sum of k copies of A.

Theorem 1.9 [6]. Let R be a simple, regular, directly finite ring which satisfies the comparability axiom.

(a) There exists a unique rank function N on R.

(b) For all x, y ε R, $xR \lesssim yR$ if and only if $N(x) \leq N(y)$.

(c) For all x, y ε R, $xR \cong yR$ if and only if $N(x) = N(y)$.

Proof. (a) Given x ε R, we define N(x) to be the supremum of all rational numbers m/n such that $m \geq 0$, $n > 0$, and $mR \lesssim n(xR)$. [6, Theorem 3.13] shows that N is a rank function, and [6, Corollary 3.15] shows that N is unique.

(b) is contained in [6, Theorem 3.13].

(c) If $N(x) = N(y)$, then in view of (b) we must have $xR \lesssim yR$ and $yR \lesssim xR$. Then $yR \cong xR \oplus A$ and $xR \cong yR \oplus B$ for some A, B, whence $xR \cong xR \oplus A \oplus B$. Since R is directly finite, so is xR, whence $A \oplus B = 0$. Thus $xR \cong yR$. ■

Corollary 1.10 [4, Corollary 4]. If R is a simple, right and left self-injective ring, there exists a unique rank function on R.

Proof. Obviously R is a nonsingular ring which is its own maximal quotient ring, hence R must be regular. According to [16, Theorems 4.7, 5.1], R is directly finite, and [3, Lemma 5', p. 832] shows that R satisfies the comparability axiom. ■

__Definition__. Let R be a regular ring and k a positive integer. Then R is said to satisfy k-__comparability__ [10] if for all $x, y \in R$, either $xR \lesssim k(yR)$ or $yR \lesssim k(xR)$. Obviously 1-comparability is just the comparability axiom, and k-comparability implies $(k + 1)$-comparability. However, $(k + 1)$-comparability does not always imply k-comparability, as shown in [10, Example 4].

Theorem 1.11 [10, Corollary 22]. Let R be a simple regular ring such that $M_n(R)$ is directly finite for all n. If R satisfies k-comparability for some k, then there exists a unique rank function on R.

2. Pseudo-rank Functions

__Definition__. A __pseudo-rank__ __function__ [4] on a regular ring R is a map $N: R \to [0, 1]$ such that

(a) $N(1) = 1$.

(b) $N(xy) \leq N(x), N(y)$.

(c) $N(e + f) = N(e) + N(f)$ for orthogonal idempotents e, f.

As a consequence of (c), we note that $N(0) = 0$. Thus a rank function is just a pseudo-rank function which assigns positive values to nonzero ring elements. As in the case of a rank function, we have $N(x + y) \leq N(x) + N(y)$ for all $x, y \in R$, and $xR \cong yR$ implies $N(x) = N(y)$.

Lemma 2.1 [4, Lemma 5]. Let R be a regular ring.

(a) If K is a proper two-sided ideal of R and N' is a rank function on R/K, then the rule $N(x) = N'(\overline{x})$ defines a pseudo-rank function N on R.

(b) If N is a pseudo-rank function on R, then the set $K = \{x \in R | N(x) = 0\}$ is a proper two-sided ideal of R, and the rule $N'(\overline{x}) = N(x)$ defines a rank function N' on R/K.

Corollary 2.2. If R is a simple regular ring, then any pseudo-rank function on R is actually a rank function.

Analogously to Proposition 1.2, we can describe the pseudo-rank functions on any semisimple Artinian ring. Thus let $R = R_1 \times \ldots \times R_t$, where each R_i is simple Artinian. Given nonnegative (rather than positive) real numbers $\alpha_1, \ldots, \alpha_t$ such that $\alpha_1 + \ldots + \alpha_t = 1$, we can define a pseudo-rank function N on R according to the rule

$$N(x_1, \ldots, x_t) = \alpha_1 \ell(x_1R_1)/\ell(R_1) + \ldots + \alpha_t \ell(x_tR_t)/\ell(R_t).$$

Conversely, every pseudo-rank function on R is of this form.

Theorem 2.3 [5]. Let R be a directly finite, regular, right self-injective ring. Then there exists a collection $\{N_M | M \in X\}$ of pseudo-rank functions on R such that for all $x, y \in R$,

(a) $xR \lesssim yR$ if and only if $N_M(x) \leq N_M(y)$ for all $M \in X$.

(b) $xR \cong yR$ if and only if $N_M(x) = N_M(y)$ for all $M \in X$.

Proof. The set $B(R)$ of all central idempotents in R forms a complete Boolean algebra [5, Proposition 4.1]. Let X be the set of all maximal ideals in $B(R)$. [This is denoted $BS(R)$ in [5].] Given $M \in X$ and $x \in R$, define $N_M(x)$ to be the infimum of all rational numbers m/n such that $m, n > 0$ and $n(xR)e \lesssim mRe$ for some $e \in B(R) - M$. According to [5, Lemma 12.3], each N_M is a pseudo-rank function on R. (a) and (b) follow from [5, Theorem 10.9 and Corollary 10.10]. ∎

<u>Definition</u>. For any regular ring R, let $\mathbb{P}(R)$ denote the set of all pseudo-rank functions on R, endowed with the relative topology induced by the product topology on $[0, 1]^R$. Given any ring map $\phi: R \to S$, we observe that $N\phi \in \mathbb{P}(R)$ for all $N \in \mathbb{P}(S)$, hence ϕ induces a map $\phi^*: \mathbb{P}(S) \to \mathbb{P}(R)$. Since ϕ^* is a restriction of the continuous map $[0, 1]^S \to [0, 1]^R$, we see that ϕ^* is continuous.

Lemma 2.4 [4, Lemma 7]. For any regular ring R, $\mathbb{P}(R)$ is a compact Hausdorff space.

Proof. Since $[0, 1]$ is a compact Hausdorff space, so is the product space $X = [0, 1]^R$, hence it suffices to show that $\mathbb{P}(R)$ is closed in X. This follows from the observation that $\mathbb{P}(R)$ is the intersection of the following closed sets: $A = \{N \in X \mid N(1) = 1\}$,

$B_{x,y} = \{N \in X \mid N(xy) \leq N(x)\}$ (for any $x, y \in R$),

$C_{x,y} = \{N \in X \mid N(xy) \leq N(y)\}$ (for any $x, y \in R$), and

$D_{e,f} = \{N \in X \mid N(e+f) = N(e) + N(f)\}$ (for any orthogonal idempotents $e, f \in R$). ∎

Theorem 2.5 [4, Theorem 8]. Let R be a direct limit (over a directed set X) of regular rings R_i. If there exists a pseudo-rank function on each R_i, then there exists a pseudo-rank function on R.

Proof. It is not hard to check that $\mathbb{P}(R)$ is the inverse limit of the

$\mathbb{P}(R_i)$, each of which is nonempty by hypothesis and compact Hausdorff by Lemma 2.4. According to [1, Chapitre I, §9.6, Proposition 8], any inverse limit of nonempty compact Hausdorff spaces is nonempty, whence $\mathbb{P}(R)$ is nonempty. ■

If R is a simple regular ring which can be expressed as a direct limit of regular rings R_i such that each R_i has a pseudo-rank function, then by Theorem 2.5 and Corollary 2.2 there must exist a rank function on R. However, if R is not simple, then there may not exist a rank function on R even if there is a rank function on each R_i, as shown by [10, Example 2].

In order to get a general existence theorem for pseudo-rank functions on a regular ring R, we turn again to the finitely generated projective R-modules, this time in the form of the group $K_0(R)$.

__Definition__. Let R be any ring. Then $K_0(R)$ is defined to be an abelian group, with generators corresponding to the members of FP(R), and with relations $A + B = C$ whenever $A \oplus B \cong C$. For $A \in FP(R)$, we use $[A]$ to denote the corresponding generator of $K_0(R)$. Given $A, B \in FP(R)$, $[A] = [B]$ if and only if A and B are __stably isomorphic__, that is, if and only if $A \oplus P \cong B \oplus P$ for some $P \in FP(R)$. Note that any element of $K_0(R)$ has the form $[A] - [B]$ for suitable $A, B \in FP(R)$.

__Definition__. A __partially ordered abelian group__ is an abelian group G equipped with a partial order $\leq$ which is translation-invariant, i.e., $x \leq y$ implies $x + z \leq y + z$. The __positive cone__ of G is the set $G^+ = \{x \in G | x \geq 0\}$. If the partial order on G is directed (upward or downward, which are equivalent), then G is called a __directed abelian group__. It is easily checked that G is directed if and only if G is generated (as a group) by G^+ [2, Proposition 3, p. 13], if and only if $G = G^+ - G^+$. A __strong unit__ in G is an element $u > 0$ such that for any $x \in G$, there is a positive integer n such that $x \leq nu$.

<u>Definition</u>. Let R be any ring. Given A, B, C, D ε FP(R), define $[A] - [B] \leqq [C] - [D]$ (in $K_0(R)$) if and only if there exists P ε FP(R) such that $A \oplus D \oplus P$ is isomorphic to a direct summand of $B \oplus C \oplus P$. (In case R is regular, this is equivalent to requiring that $A \oplus D \oplus P \lesssim B \oplus C \oplus P$.) This defines a relation $\leqq$ on $K_0(R)$ which is clearly reflexive and transitive. The basic properties of this relation, as expressed in the following proposition, are easily checked.

Proposition 2.6. [18, Proposition 2.1]. Let R be a nonzero ring such that $M_n(R)$ is directly finite for all n.

(a) $(K_0(R), \leqq)$ is a directed abelian group.

(b) $K_0(R)^+ = \{[A] \mid A \in FP(R)\}$.

(c) [R] is a strong unit in $K_0(R)$.

<u>Definition</u>. Let G be a partially ordered abelian group with a strong unit u. If f is a monotone (i.e., order-preserving) homomorphism of $G \to \mathbb{R}$ such that $f(u) = 1$, then we shall refer to f as a <u>pseudo-rank function on</u> (G, u).

Proposition 2.7 [18, Proposition 2.4]. Let R be a regular ring such that $M_n(R)$ is directly finite for all n. If f is a pseudo-rank function on $(K_0(R), [R])$, then the rule $N_f(x) = f([xR])$ defines a pseudo-rank function N_f on R. Conversely, if N is any pseudo-rank function on R, then there exists a unique pseudo-rank function f on $(K_0(R), [R])$ such that $N_f = N$.

Proof. Analogous to Proposition 1.4. ∎

Theorem 2.8 [18, Corollary 3.3]. If G is a directed abelian group with a strong unit u, then there exists a pseudo-rank function on (G,u).

Corollary 2.9. Let R be a regular ring. Then there exists a pseudo-rank function on R if and only if R has a proper two-sided ideal K such that $M_n(R/K)$ is directly finite for all n.

Proof. If there exists a pseudo-rank function on R, then by Lemma 2.1 R has a proper two-sided ideal K such that there is a rank function on R/K. According to Lemma 1.6, $M_n(R/K)$ is directly finite for all n.

Conversely, assume that R has a proper two-sided ideal K such that $M_n(R/K)$ is directly finite for all n. Then Lemma 2.6 shows that $K_0(R/K)$ is a directed abelian group with strong unit [R/K]. By Theorem 2.8 and Proposition 2.7, there is a pseudo-rank function on R/K, and consequently there is a pseudo-rank function on R. ■

Corollary 2.10. Let R be a simple regular ring. Then there exists a rank function on R if and only if $M_n(R)$ is directly finite for all n.

Proof. Corollaries 2.2 and 2.9. ■

<u>Definition</u>. A ring R is <u>unit-regular</u> if for each $x \in R$ there is a unit $u \in R$ such that $xux = x$. It is easy to check that any commutative regular ring is unit-regular. More generally, any regular ring which is <u>abelian</u> (i.e., all idempotents are central) is unit-regular. As another example, any direct limit of semisimple artinian rings is unit-regular.

Proposition 2.11 [9, Theorem 2]. A regular ring R is unit-regular if and only if for all A, B, C $\in$ FP(R), $A \oplus B \cong A \oplus C$ implies $B \cong C$.

Corollary 2.12 [11, Corollary 7]. Unit-regularity is Morita-invariant.

Proof. If R and S are Morita-equivalent rings, then it is well-known that regularity carries over from R to S. Since FP(R) and FP(S) are equivalent categories, Proposition 2.11 shows that unit-regularity carries over as well. ■

Corollary 2.13. If R is a unit-regular ring, then $M_n(R)$ is directly finite for all n.

Proof. In view of Proposition 2.11, we see that nR is a directly finite module for all n. ■

The converse of Corollary 2.13 fails, as shown by an example of Bergman [9, Example 2]. However, it is not known whether there exist simple regular directly finite rings which are not unit-regular.

Corollary 2.14 [6, Theorem 3.9 and Corollary 3.10]. If R is a directly finite regular ring which satisfies the comparability axiom, then R is unit-regular.

Corollary 2.15 [5]. If R is a directly finite, regular, right self-injective ring, then R is unit-regular.

Proof. Since R is regular and right self-injective, we infer that any A, B, C $\in$ FP(R) are nonsingular injective right R-modules. Since R is directly finite, it follows from [5, Theorem 3.6] that A is directly finite. Then $A \oplus B \cong A \oplus C$ implies $B \cong C$ by [5, Theorem 3.8]. ■

Theorem 2.16 [18, Corollary 3.5]. If R is a nonzero unit-regular ring, then there exists a pseudo-rank function on R.

Proof. Corollaries 2.9 and 2.13. ■

Corollary 2.17. If R is a simple unit-regular ring, then there exists a rank function on R.

Proof. Theorem 2.16 and Corollary 2.2. ■

§3. Completions.

<u>Definition</u>. Let R be a regular ring with a rank function N. Then the rule $\delta(x,y) = N(x - y)$ defines a metric δ on R [17, p. 231], called the <u>rank-metric associated with</u> N, or the N-<u>metric</u> for short. As the following lemma shows, addition, multiplication, and N are all uniformly continuous with respect to δ.

Lemma 3.1 [17, p. 232]. Let R be a regular ring with a rank function N and associated rank-metric δ.

(a) $\delta(x + y, z + w) \leq \delta(x, z) + \delta(y, w)$ for all $x, y, z, w \in R$.

(b) $\delta(xy, zw) \leq \delta(x, z) + \delta(y, w)$ for all $x, y, z, w \in R$.

(c) $|N(x) - N(y)| \leq \delta(x, y)$ for all $x, y \in R$.

Theorem 3.2 [7, Theorem 3.7]. Let R be a regular ring with a rank function N, and let R^* be the completion of R in the N-metric. Then R^* is a regular ring, N extends continuously to a rank function N^* on R^*, and R^* is complete in the N^*-metric.

Theorem 3.3 [4, Theorem 14]. Let R be a regular ring with a rank function N. If R is complete in the N-metric, then R is right and left self-injective.

Proof. Let J be a right ideal of R, f any homomorphism of J into R_R. Since there is a rank function on R, R does not contain any uncountable direct sums of nonzero right ideals. As a result, there exist orthogonal idempotents $e_1, e_2, \ldots \in R$ such that $\oplus e_nR$ is essential in J [4, Lemma 13]. Set $x_n = fe_n$ for all n.

It is not hard to check that the partial sums of the series $\sum x_n$ form a Cauchy sequence with respect to the N-metric. By completeness, this series must converge to some $x \in R$. Then $xe_n = fe_n$ for all n, hence left multiplication by x agrees with f on $\oplus e_nR$. Inasmuch as $J/(\oplus e_nR)$ is a singular right R-module while R_R is nonsingular, it follows that f must be given by left multiplication by x.

Thus R_R is injective, and by symmetry ${}_RR$ is injective also. ∎

Corollary 3.4. Let R be a regular ring with a rank function N. If R^* is the completion of R in the N-metric, then R^* is a regular, right and left self-injective ring.

Proof. Theorems 3.2 and 3.3. ∎

Definition. A convex combination of vectors $v_1, \ldots, v_t$ in a real vector space V is any linear combination $\alpha_1v_1 +\ldots+ \alpha_tv_t$ such that all $\alpha_i \geq 0$ and $\alpha_1 +\ldots+ \alpha_t = 1$. If all $\alpha_i > 0$, then we refer to $\alpha_1v_1 +\ldots+ \alpha_tv_t$ as a positive convex combination of $v_1, \ldots, v_t$. In case V is a topological vector space, it is also possible to talk about infinite convex combinations in V. Thus suppose we have $v_1, v_2, \ldots \in V$ and nonnegative real numbers $\alpha_1, \alpha_2, \ldots$ such that $\sum\alpha_i = 1$. If the series $\sum\alpha_iv_i$ converges to some $v \in V$, then v is called a convex combination of the v_i. If in addition all $\alpha_i > 0$, then v is a positive convex combination of the v_i.

For any regular ring R, $\mathbb{P}(R)$ is a subset of the real vector space $\mathbb{R}^R$, and we observe that any convex combination of pseudo-rank functions in $\mathbb{P}(R)$ is again a pseudo-rank function, i.e., $\mathbb{P}(R)$ is a convex subset of $\mathbb{R}^R$. In addition, $\mathbb{P}(R)$ is closed under infinite convex combinations, as follows. Suppose that $N_1, N_2, \ldots \in \mathbb{P}(R)$, that $\alpha_1, \alpha_2, \ldots$ are nonnegative real numbers, and that $\sum\alpha_i = 1$. The series $\sum\alpha_iN_i(x)$ converges for every $x \in R$, hence we may define a map $N: R \to [0, 1]$ by the rule $N(x) = \sum\alpha_iN_i(x)$. The partial sums of the series $\sum\alpha_iN_i$ converge pointwise to N, so that $\sum\alpha_iN_i$ converges to N in the product topology on $\mathbb{R}^R$. We write $N = \sum\alpha_iN_i$, and we check that $\sum\alpha_iN_i \in \mathbb{P}(R)$.

Definition. Let A be a convex subset of a real vector space. An extreme point of A is any element of A which cannot be expressed as a positive convex combination of distinct elements of A. In other words, an element $x \in A$ is an extreme point of A if and only if $x = \alpha y + (1 - \alpha)z$ [with $0 \leq \alpha \leq 1$ and $y, z \in A$] occurs only when $\alpha = 0, 1$ or $y = z$.

Proposition 3.5 [4, Proposition 17]. If R is a regular ring, then $\mathbb{P}(R)$ is the closure of the convex hull of its extreme points.

Proof. The product space $\mathbb{R}^R$ is a topological vector space whose dual separates points, and we have seen that $\mathbb{P}(R)$ is a compact convex subset of $\mathbb{R}^R$. The proposition thus follows from the Krein-Milman Theorem [14, Theorem 3.21]. ∎

Definition. If P is a pseudo-rank function on a regular ring R, we define the kernel of P to be the set $\ker(P) = \{x \in R | P(x) = 0\}$, which by Lemma 2.1 is a proper two-sided ideal of R.

Proposition 3.6 [4]. Let R be a regular ring with a rank function N, and let R^* be the completion of R in the N-metric. Let P be a pseudo-rank function on R, and assume that P is uniformly continuous in the N-metric.

(a) P extends uniquely to a uniformly continuous pseudo-rank function P^* on R^*.

(b) P is an extreme point of $\mathbb{P}(R)$ if and only if $\ker(P^*)$ is a maximal two-sided ideal of R^*.

Proof. (a) is contained in [4, Lemma 16].

(b) Since $\ker(P^*)$ is closed in the N^*-topology, it follows from the completeness of R^* that $\ker(P^*)$ is generated by a central idempotent [4, Proposition 18]. In view of Corollary 3.4, it follows that $R^*/\ker(P^*)$ is a regular, right and left self-injective ring. According to [16, Theorems 4.7, 5.1], $R^*/\ker(P^*)$ is directly finite. It now follows from [13, Proposition 2.7] that $\ker(P^*)$ is a maximal two-sided ideal of R^* if and only if $R^*/\ker(P^*)$ is an indecomposable ring.

Thus if $\ker(P^*)$ is not a maximal two-sided ideal of R^*, $R^*/\ker(P^*)$ must be a direct product of two nonzero rings. Since P^* induces a rank function on $R^*/\ker(P^*)$ by Lemma 2.1, we see using Lemma 1.1 that $P^* = \alpha P_1^* + (1-\alpha)P_2^*$ with $0 < \alpha < 1$ and distinct $P_1^*, P_2^* \in \mathbb{P}(R^*)$. Restricting these maps to R, we obtain $P = \alpha P_1 + (1-\alpha)P_2$ with $P_1, P_2 \in \mathbb{P}(R)$. Since P_1^* and P_2^* are uniformly continuous [4, Lemma 16], we must have $P_1 \neq P_2$, hence P is not an extreme point of $\mathbb{P}(R)$.

If P is not an extreme point of $\mathbb{P}(R)$, then we must have $P = \alpha P_1 + (1-\alpha)P_2$ for $0 < \alpha < 1$ and distinct $P_1, P_2 \in \mathbb{P}(R)$. By [4, Lemma 16], each P_i extends to a uniformly continuous $P_i^* \in \mathbb{P}(R^*)$, and $P^* = \alpha P_1^* + (1-\alpha)P_2^*$. Observing that $\ker(P^*) \subsetneq \ker(P_i^*)$ for each i, we infer that P_1^* and P_2^* induce distinct pseudo-rank functions on $R^*/\ker(P^*)$. In view of Corollaries 1.10 and 2.2, we conclude that $R^*/\ker(P^*)$ is not a simple ring. ■

Corollary 3.7 [4, Corollary 20]. Let R be a regular ring with a rank function N, and let R^* be the completion of R in the N-metric. Then R^* is a simple ring if and only if N is an extreme point of $\mathbb{P}(R)$.

Proof. By Theorem 3.2, N extends continuously to a rank function N^* on R^*. Since $\ker(N^*) = 0$, the corollary follows immediately from Proposition 3.6. ■

Corollary 3.8 [4, Corollary 21]. Let R be a regular ring with a rank function N, and let R^* be the completion of R in the N-metric.

(a) If R has exactly one rank function, then R^* is a simple ring.

(b) If R is a simple ring, then N can be chosen so that R^* is a simple ring.

(c) If R has more than one rank function, then N can be chosen so that R^* is not a simple ring.

Proof. (a) Given any $P \in \mathbb{P}(R)$, we see that $(1/2)N + (1/2)P$ is a rank function on R. Then $(1/2)N + (1/2)P = N$ and so $P = N$. Thus $\mathbb{P}(R) = \{N\}$, whence N is an extreme point of $\mathbb{P}(R)$.

(b) According to Proposition 3.5, there exists an extreme point $P \in \mathbb{P}(R)$. Since P is a rank function by Corollary 2.2, we may choose $N = P$.

(c) If N_1, N_2 are distinct rank functions on R, then $N = (1/2)N_1 + (1/2)N_2$ is a rank function on R which is not an extreme point of $\mathbb{P}(R)$. ∎

Theorem 3.9 [4, Theorem 19]. Let R be a regular ring with a rank function N, let R^* be the completion of R in the N-metric, and let k be a positive integer. Then R^* is a direct product of k simple rings if and only if N is a positive convex combination of k distinct extreme points in $\mathbb{P}(R)$.

Proof. First assume that $R^* = R_1 \times \ldots \times R_k$ with each R_i simple, and let e_i denote the unit of R_i. Proceeding as in Lemma 1.1, we must have $N^* = \alpha_1 P_1^* + \ldots + \alpha_k P_k^*$ for suitable positive real numbers α_i such that $\alpha_1 + \ldots + \alpha_k = 1$ and suitable $P_i^* \in \mathbb{P}(R^*)$ such that $P_i^*(e_i) = 1$. Restricting these maps to R, we get $N = \alpha_1 P_1 + \ldots + \alpha_k P_k$ with each $P_i \in \mathbb{P}(R)$. Since the P_i^* are uniformly continuous [4, Lemma 16], we see that the P_i are distinct. Also, Proposition 3.6 shows that the P_i are extreme points of $\mathbb{P}(R)$.

Conversely, assume that $N = \alpha_1 P_1 + \ldots + \alpha_k P_k$ for suitable positive real numbers α_i such that $\alpha_1 + \ldots + \alpha_k = 1$ and suitable distinct extreme points $P_i \in \mathbb{P}(R)$. By [4, Lemma 16], each P_i extends to a uniformly continuous pseudo-rank function $P_i^* \in \mathbb{P}(R^*)$, and

$N^* = \alpha_1 P_1^* + \ldots + \alpha_k P_k^*$. Using Proposition 3.6 and Corollary 1.10, we infer that the kernels of the P_i^* are distinct maximal two-sided ideals of R^*. Inasmuch as $\cap \ker(P_i^*) = \ker(N^*) = 0$, we conclude that $R^* \cong [R^*/\ker(P_1^*)] \times \ldots \times [R^*/\ker(P_k^*)]$ is a direct product of k simple rings. ∎

Theorem 3.10 [4, Theorem 22]. Let R be a regular ring with a rank function N, and let R^* be the completion of R in the N-metric. Then R^* is an infinite direct product of simple rings if and only if N is a positive convex combination of infinitely many distinct extreme points in $\mathbb{P}(R)$.

The possibilities indicated in Theorems 3.9 and 3.10 can all occur for the same ring (with different rank functions). For instance, [4, Example C] is a simple regular ring R which possesses rank functions $N_0, N_1, N_2, \ldots, N_\infty$ such that: (a) the completion of R in the N_0-metric has no simple ring direct summands; (b) for k = 1, 2, ..., the completion of R in the N_k-metric is a direct product of exactly k simple rings; and (c) the completion of R in the N_∞-metric is a direct product of infinitely many simple rings.

4. Open questions.

Let R be a regular ring.

1. If A, B $\in$ FP(R) are directly finite, is $A \oplus B$ directly finite? [This is true if R is self-injective on either side, but it can fail for non-regular rings.]

2. If R is directly finite and S is Morita-equivalent to R, is S directly finite? [This can fail if R is not regular.] Perhaps this is true if $M_2(R)$ is directly finite.

3. If R is directly finite and M is a maximal two-sided ideal of R, is R/M directly finite? [This can fail for prime ideals.]

4. [Roos' Conjecture] If R is directly finite and right self-injective, is R also left self-injective? [This is true in the "Type I" case.]

5. If R is simple and directly finite, is R unit-regular? [This can fail if R is not simple.]

6. Let R be unit-regular, A, B ε FP(R), n a positive integer. If nA $\cong$ nB, is A $\cong$ B? [I.e., is $K_0(R)$ torsion-free?] If nA $\lesssim$ nB, is A $\lesssim$ B? [The answer to both questions is yes if R is self-injective on either side.]

7. If there exists a rank function on R, is R unit-regular?

References

1. N. Bourbaki, Topologie Générale (Eléménts de Mathématique, Livre III), Paris (1965) Hermann.

2. L. Fuchs, Partially Ordered Algebraic Systems, Oxford (1963), Pergamon Press.

3. K. R. Goodearl, "Prime ideals in regular self-injective rings," Canadian J. Math. 25(1973) 829-839.

4. ________, "Simple regular rings and rank functions" Math. Annalen 214(1975), 267-287.

5. ________ and A. K. Boyle, "Dimension theory for nonsingular injective modules," (to appear).

6. ________ and D. Handelman, "Simple self-injective rings," Communications in Algebra 3(1975), 797-834.

7. I. Halperin, "Regular rank rings," Canadian J. Math. 17(1965), 709-719.

8. ________, "Extension of the rank function," Studia Math. 27(1966), 325-335.

9. D. Handelman, "Perspectivity and cancellation in regular ring ," J. Algebra (to appear).

10. ________, "Simple regular rings with a unique rank function," J. Algebra (to appear).

11. M. Henriksen, "On a class of regular rings that are elementary divisor rings," Arch. der Math. 24(1973), 133-141.

12. I. Kaplansky, "Projective modules," Ann. of Math. 68(1958), 372-377.

13. G. Renault, "Anneaux réguliers auto-injectifs à droite," Bull. Soc. Math. France 101(1973), 237-254.

14. W. Rudin, Functional Analysis, New York (1973) McGraw-Hill.

15. J. C. Shepherdson, "Inverses and zero divisors in matrix rings," Proc. London Math. Soc. 1(1951) 71-85.

16. Y. Utumi, "On continuous rings and self-injective rings," Trans. American Math. Soc. 118(1965), 158-173.

17. J. von Neumann, Continuous Geometry, Princeton (1960) Princeton University Press.

18. K. R. Goodearl and D. Handelman, "Rank functions and K_o of regular rings," J. Pure Applied Algebra (to appear).

SOME ASPECTS OF NONCOMMUTATIVE NOETHERIAN RINGS

Robert Gordon*
Temple University
Philadelphia, Pennsylvania

Given a module M with Krull dimension there is a notion of a partially ordered set spec M, called the <u>critical spectrum of M</u>, defined in terms of certain equivalence classes of critical modules. The paper starts with a discussion of this notion and a description of the minimal elements of spec M. Later on the notation is justified by showing that spec M, for M a finitely generated module over a right FBN ring (see Section 2), can be identified with the poset of prime ideals of the ring containing the annihilator of M.

We also study primary modules and decomposition, and the interrelationship of these concepts with the critical spectrum. Our results in Section 2 show that the theory of primary decomposition for finitely generated modules over a right FBN ring echoes the commutative theory; and yet, paradoxically, finitely generated primary modules themselves can exhibit pathology antithetic to the commutative behavior. Most of this is taken from [4]. However the results, in Section 3, concerning the existence of invariant primary submodules of a finitely generated module over an FBN ring, are new.

Later in the same section we reformulate our result [3], that an FBN Macaulay ring is an order in an Artinian ring, as a consequence of more general results concerning localization in Noetherian rings. We also prove that a finitely generated module over an FBN order in an Artinian ring is torsionfree if and only if each of its critical composition factors is a critical composition factor of the ring. In this

---

*Research partially supported by N.S.F. Grant No. MPS 75-06327.

case the critical composition factors of the module are precisely the minimal elements of its critical spectrum.

Throughout this paper we take for granted the relevant notions and results of [6]. Rings are assumed to have an identity, modules are unitary right modules, and the Krull dimension of a module M is denoted by K dim M.

I would like to express my thanks to the organizers of the conference.

§1. The Critical Spectrum

If M is a module and C a critical module we say that M is C-torsion if it is torsion in the torsion theory cogenerated by C; i.e., Hom (M, E(C)) = 0, E(C) the injective envelope of C. We note that M is C-torsion if and only if no subfactor module of M is isomorphic to a nonzero submodule of C. We say that M is C-torsionfree if its largest C-torsion submodule, $T_C(M)$, is 0.

When it is permissible we will not hesitate to identify C with the equivalence class of critical modules with injective envelope isomorphic to E(C). We do recall, though, that two critical modules have isomorphic injective envelopes if and only if each has a nonzero submodule isomorphic to a submodule of the other.

We partially order the set of equivalence classes of critical modules by defining the class of C_1 to be less than or equal to the class of C_2 is C_1 is C_2-torsionfree. We then define spec M to be the subposet determined by critical modules C such that M is not C-torsion. The elements of the subposet ass M of spec M consisting of equivalence classes of critical submodules of M are termed associated critical modules of M.

We call M primary if ass M consists of a single element and use the term C-primary when that element is C. We also refer to primary submodules of M--meaning submodules M' such that M/M' is primary. C-primary submodules are defined analogously.

The notion of spec M can be reformulated in terms of primary submodules of M.

Lemma 1.1. Let C be a critical module.

(i) $C \in$ spec M if and only if M has a C-primary submodule.

(ii) $T_C(M)$ is the intersection of the C-primary submodules of M.

Proof. It clearly suffices to prove (ii), so let D be the intersection specified. Then $T_C(M) \subseteq D$. But also

$$T_C(M) = \cap\{\ker f \mid f:M \to E(C)\}$$

and $M/\ker f$, where $f:M \to E(C)$, is C-primary. Thus $D \subseteq T_C(M)$. ∎

The next result exposes the significance of the minimal elements of spec M.

Theorem 1.2. If $C \in$ spec M and M has Krull dimension then the following statements are equivalent.

(i) C is a minimal element of spec M.

(ii) Every nonzero C-torsionfree homomorphic image of M is C-primary.

(iii) M has d.c.c. on submodules M' such that M/M' is C-torsionfree.

(iv) M has no infinite descending chain of C-primary submodules.

Proof. (i) ⇒ (ii). Clear.

(ii) ⇒ (iii). Let $M = M_0 \supseteq M_1 \supseteq \ldots$ be a descending chain of submodules M_i of M such that M/M_i is C-torsionfree. Without loss of generality the intersection of the chain is 0 and then, by hypothesis, M is C-primary. Thus M has an essential submodule E of Krull dimension α where $\alpha = K \dim C$.

Now each factor in the chain $E = E \cap M_0 \supseteq E \cap M_1 \supseteq \ldots$ is C-torsionfree since each is isomorphic to a submodule of some M/M_i. But any nonzero C-torsionfree module must have Krull dimension $\geq \alpha$. Thus, for some n, $E \cap M_n = E \cap M_{n+1} = \ldots$ and so $M_n = 0$.

(iii) ⇒ (iv). Trivial.

(iv) ⇒ (i). Suppose that C is not minimal. Then M has a critical C-torsionfree subfactor module Y/X which is not in the equivalence class of C. Choose a proper factor module Y/A_1 of Y/X such that Y/A_1 is not C-torsion. By Lemma 1.1 M/A_1 has a C-primary submodule, N_1/A_1 say. Next choose a proper factor A_1/A_2 of A_1/X such that A_1/A_2 is not C-torsion. Thus N_1/A_2 has a C-primary submodule N_2/A_2.

This construction yields an infinite descending chain $M = N_0 \supset N_1 \supset \ldots$ of submodules N_i of M such that N_i/N_{i+1} is C-primary for all i. But an extension of a C-primary module by a C-primary module is C-primary. Hence each N_i is a C-primary submodule of M. ■

If $C \in$ ass M is a minimal element of spec M then we say that C is an <u>isolated critical module of M</u>. An associated critical module of M which is not isolated we call <u>embedded</u>. In this terminology we have

<u>Corollary 1.3</u>. If M is Noetherian then M is primary and its associated critical module is isolated if and only if there is a critical module C and a chain $0 = M_0 \subset M_1 \subset \ldots \subset M_m = M$ of submodules M_i of M such that, for each i, M_i/M_{i-1} is isomorphic to an essential extension of a nonzero submodule of C by a C-torsion module.

<u>Proof</u>. => . Let M be C-primary, C isolated. We assume that C itself is a submodule of M. Let M_1/C, M_1 a submodule of M containing C, be the C-torsion submodule of M/C. Since M_1/C is torsion and M_1 is torsion-free M_1 is an essential extension of C.

Now $T_C(M/M_1) = 0$ by construction and so, if $M \neq M_1$, then M/M_1 is C-primary. The fact that C is an isolated critical module of M/M_1 makes the rest of the proof clear.

<=. Since every proper factor module of C must be C-torsion it is plain that each factor M_i/M_{i-1} in the specified chain has d.c.c. on C-primary submodules. Thus M has d.c.c. on C-primary submodules. But, because each M_i/M_{i-1} is C-torsionfree, M is C-torsionfree. ■

We hasten to point out that the associated critical module of a primary Noetherian module need not be isolated (see the example below Corollary 2.8). In particular a Noetherian module need not have an isolated critical module. However it is fairly clear that spec M, for M Noetherian, has minimal elements.

We turn next to primary decomposition. For this let M be a non-zero module with Krull dimension. Since M has finite uniform dimension ass M is a finite set with distinct members $C_1, \ldots, C_r$ say. A <u>primary decomposition</u> of M is a collection of C_i-primary submodules M_i of M such that $\bigcap_{i=1}^{r} M_i = 0$. We call M_i a <u>C_i -primary component</u> of M.

To see that M has a primary decomposition note that

$$E(M) \simeq E(C_1)^{n_1} \oplus \ldots \oplus E(C_r)^{n_r}$$

for some integers n_i. Then a suitable choice of a C_i-primary component of M is the kernel of the obvious map $M \to E(C_i)^{n_i}$.

We remark that this primary decomposition is finer than the one studied in [4]. However it coincides with the other in many special cases, for example, when the underlying ring is a right FBN ring--see Section 2. Both the notion of primary decomposition discussed here and the following result are due to Gabriel [1].

<u>Lemma 1.4</u>. Let $M_1, \ldots, M_r$ be a finite collection of primary submodules of a module M with Krull dimension such that, for $i \neq j$, M/M_i and M/M_j have different associated critical modules. If $\bigcap M_i = 0$ then this intersection is irredundant if and only if it is a primary decomposition of M.

<u>Proof</u>. If C is a critical submodule of M then, since $\bigcap M_i = 0$, there is a submodule C' of C such that $C' \cap M_i = 0$ for some i. Thus ass $M \subseteq \bigcup$ ass M/M_i. The same argument shows that if the intersection

$\cap M_i = 0$ is a primary decomposition then it is irredundant.

Conversely suppose that the intersection is irredundant. Then for each i some nonzero submodule of M/M_i embeds in $\cap_{j \neq i} M_j \subseteq M$. Thus $\cup$ ass $M/M_i \subseteq$ ass M. ■

The next basic, and easy, result underscores the importance of understanding spec M when M is primary.

Lemma 1.5. If $\cap M_i = 0$ is a primary decomposition of module M with Krull dimension then spec M = $\cup$ spec M/M_i.

We will prove, in Section 3, that a finitely generated module M over an FBN ring has a primary decomposition such that each primary component is an invariant submodule of M. We would like to prove a similar result for modules over more general classes of rings. (Indeed, for the rings themselves.) Our next result can be interpreted as limited success. Its proof is an immediate consequence of 1.2.

Corollary 1.6. If M is a module with Krull dimension then every associated critical module of M is isolated if and only if

$$\bigcap_{C \in \text{ass } M} T_C(M) = 0$$

is a primary decomposition of M and C is an isolated critical module of $M/T_C(M)$ for each C ε ass M.

We should point out that $T_C(M)$ is a C-primary component of M contained in every C-primary submodule of M whenever C is an isolated critical module of a module M with Krull dimension. Yet even if every associated critical module is isolated there can be more than one primary decomposition (see [5, Example 4.8]). In fact there can be infinitely many in which every primary component is an invariant submodule.

The modules characterized in Corollary 1.6 contain a large and useful class. We say that a module is α-Macaulay if it is nonzero and each of its nonzero finitely generated submodules has Krull dimension α. A Macaulay module is a module that is α-Macaulay for some α. By Theorem 1.2 every associated critical module of a Macaulay module with Krull dimension is isolated. (This also uses [6, Theorem 4.1].) We should caution the reader that the term "Macaulay" is used in [4] with a somewhat different meaning.

Lemma 1.7. If M is a faithful α-Macaulay module over a ring R of Krull dimension α then, for some n, R is isomorphic to a submodule of the direct sum of n copies of M.

Proof. Choose $0 \neq m_1 \in M$ and let $B_1 = \operatorname{ann} m_1$. Suppose that $B_1 \neq 0$. Then, using the faithfulness of M, choose $m_2 \in M$ such that $B_1 \not\subseteq \operatorname{ann} m_2$. Thus $B_2 = \operatorname{ann} \{m_1, m_2\} \subset B_1$. Continuing in this manner construct a proper descending chain $B_1 \supset B_2 \supset \dots$ of right ideals B_i such that, for each i, R/B_i is isomorphic to a submodule of M^i. Since K dim R = α it follows that the chain reaches 0. ∎

In the equivalence class of any critical module there is a cyclic module. Thus spec M ⊆ spec R for any R-module M.

Corollary 1.8. Using the hypothesis and notation of Lemma 1.7 we have spec M = spec R.

To further exploit 1.7 we must remind the reader that a critical composition series of a Noetherian module M is a chain $0 = M_0 \subset M_1 \subset \dots \subset M_n = M$ of submodules M_i such that M_i/M_{i-1} is α_i-critical and $\alpha_1 \leq \alpha_2 \leq \dots \leq \alpha_n$. By [2, Corollary 2.8], the injective envelopes of the critical composition factors M_i/M_{i-1} are determined by M up to order and isomorphism. We denote the finite partially ordered subset of spec M consisting of equivalence classes of critical composition factors of M by comp M. We remark that M is α-Macaulay if and only if the elements of comp M are all α-critical. We also require

Lemma 1.9. If N is a submodule of a Noetherian module M then comp N $\subseteq$ comp M.

Proof. Let $0 = N_0 \subset N_1 \subset \ldots \subset N_r = N$ be a critical composition series of N where N_i/N_{i-1} is, say, α_i-critical. Put $N_r' = M$ and, for $i < r$, let N_i'/N_i be the largest submodule of M/N_i of Krull dimension $< \alpha_{i+1}$. It follows that for each i we have the picture

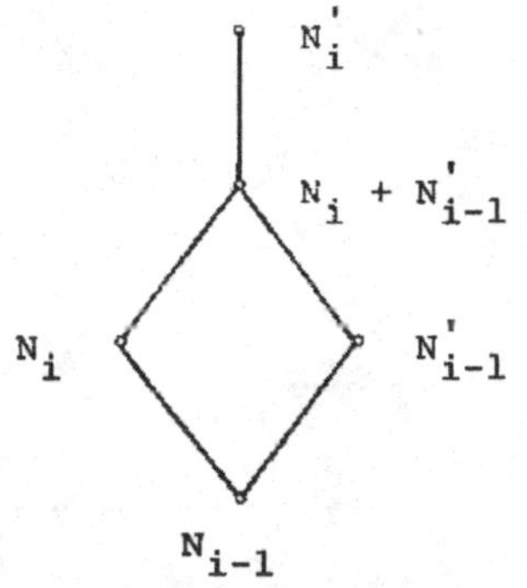

This makes it plain that K dim $N_i'/N_{i-1}' \leq \alpha_{i+1}$. But, by construction, every nonzero submodule of N_{i+1}'/N_i' has Krull dimension $\geq \alpha_{i+1}$. Thus the chain $0 \subseteq N_0' \subset N_1' \subset \ldots \subset N_r' = M$ can be refined to a critical composition series. Moreover the picture shows that some nonzero submodule of N_i/N_{i-1} is embedded in a critical composition factor of N_i'/N_{i-1}'. ∎

Of course this implies that ass M $\subseteq$ comp M.

Corollary 1.10. If M is a faithful finitely generated α-Macaulay module over a right Noetherian ring R of Krull dimension α then comp R = comp M.

<u>Proof</u>. It suffices to show that comp M $\subsetneq$ comp R. But it is easy to see that R itself is α-Macaulay. Thus if R/I, I a right ideal of R, is α-critical then I is 0 or α-Macaulay. In either case the chain $0 \subsetneq I \subset R$ can be refined to a critical composition series of R. ∎

§2. RIGHT FBN RINGS

For any ring R with Krull dimension there is an obvious surjective map from spec R to the set of prime ideals of R. This map is defined by sending critical modules C to their assassinators ass C, where ass C is the set of elements of R which kill nonzero submodules of C.

A further property of this map is that it is order preserving. For let C and D be critical modules such that C is D-torsionfree and let P and Q be the respective assassinators of C and D. Since Hom (C', E(D)) $\neq$ 0 for some submodule C' of C with annihilator P, E(D) has a nonzero submodule killed by P. Thus P $\subsetneq$ Q.

We recall that a right FBN ring is a right Noetherian ring such that the map described above is injective. (Right FBN rings are characterized amongst right Noetherian rings by their prime factor rings being right bounded--FBN is an abbreviation for "fully bounded Noetherian".) We also recall that an FBN ring is a ring which is a left FBN ring as well as a right FBN ring.

Proposition 2.1. If R is a right FBN ring then spec R and the set of prime ideals of R are canonically isomorphic posets.

Proof. Let P $\subsetneq$ Q be prime ideals of R and let C and D be critical modules such that ass C = P, ass D = Q. We must show that C is D-torsionfree. For this we can assume that C is a right ideal of R/P.

If C is not D-torsionfree then, since C embeds in each of its nonzero submodules, C is D-torsion. But then R/P has an essential D-torsion right ideal. This makes R/P itself D-torsion, by Goldie's Theorem. But Hom (R/P, E(R/Q)) $\neq$ 0 and yet E(R/Q) is a direct sum of copies of E(D). ■

Henceforth when dealing with right FBN rings we freely identify elements of spec R with prime ideals P. This makes terms such as

"P-torsionfree" or "P-primary", and notation such as $T_P(R)$, self-explanatory.

The next theorem shows that, for a finitely generated module M over a right FBN ring, spec M can be described as the poset of prime ideals containing ann M. In particular an associated prime of M is isolated if and only if it is minimal amongst primes lying over ann M. There is an example [4, Example 4.2] of an FBN ring R, finitely generated as a module over its center, and having a minimal associated prime P which is not isolated, i.e., P is not a minimal prime of R. We shall see, in Corollary 3.3, that every P-primary right ideal of R contains an infinite descending chain of P-primary ideals; and $T_P(R)$ certainly cannot be a P-primary component. Thus the usual definition of an "isolated" prime as being minimal amongst associated primes does not seem suitable to us even in mildly noncommutative situations.

We will require the basic fact [7] that the Krull dimension of a finitely generated faithful module over a right FBN ring is the same as the Krull dimension of the ring. If M is a module with Krull dimension we will denote by τM its largest submodule of Krull dimension minimal amongst Krull dimensions of nonzero submodules.

<u>Theorem 2.2</u>. If M is a finitely generated faithful module over a right FBN ring R then spec M = spec R and comp R $\subseteq$ comp M.

<u>Proof</u>. We prove that comp R $\subseteq$ comp M. The proof of the other relation, using Corollary 1.8, is slightly easier.

If M is a critical module then the result is a consequence of Corollary 1.10. Thus we can assume the result valid for modules of critical composition series length smaller than that of M.

Now let $L = \tau M$, $\alpha = K \dim L$ and $A = \operatorname{ann} M/L$. By induction comp R/A $\subseteq$ comp M/L. Also MA, being a nonzero submodule of L, is α-Macaulay. Therefore, by 1.10, comp R/B = comp MA where B = ann MA.

Next note that, since M is faithful, B = r(A). In particular K dim A = K dim R/B = K dim MA = α. But $m\tau(R) \neq 0$ for some $m \in M$. Thus $\tau R \geq \alpha$ follows and so A is α-Macaulay. Hence comp A = comp R/B, by 1.10.

Using Lemma 1.9 and combining we get comp A $\subseteq$ comp L. Thus comp R = comp A $\cup$ comp R/A $\subseteq$ comp L $\cup$ comp M/L = comp M. This last follows because neither R/A nor M/L has a critical composition factor of dimension α or less. ∎

The method of this proof is apparently not quite sufficient to show that ass R $\subseteq$ ass M, and we resort to the classical method of primary decomposition for this.

<u>Theorem 2.3</u>. A right FBN ring with a faithful finitely generated P-primary module is P-primary.

<u>Proof</u>. Let R be the ring, M the module, and let C be a critical right ideal such that ass C = ann C = Q say. Then, since M is faithful, MC is a finitely generated faithful module over the prime right FBN ring R/Q.

Since M is primary τM is essential in M. There $\tau(MC) = MC \cap \tau M$ is essential in MC. It follows that K dim $MC/\tau(MC)$ < K dim R/Q. But K dim R/Q = K dim MC. Thus $\tau(MC)$ = MC and so MC $\subseteq \tau M$. But then C is contained in the annihilator A of $M/\tau M$; and in the proof of 2.2 we saw that A is α-Macaulay, where α = K dim τM. So C is α-critical.

To finish choose an element $m \in M$ such that $0 \neq mC \subseteq \tau M$. Then $C \simeq mC$ since K dim mC = α. ∎

<u>Corollary 2.4</u>. A right FBN ring is primary if and only if it has a finitely generated faithful uniform module.

<u>Proof</u>. If the ring, R say, is primary then $R \subseteq I^n$ for some indecomposable injective module I. Let $U = U_1 + \ldots + U_n$ where

$U_i = \text{im}(R \to I^n \xrightarrow{\pi_i} I)$ and π_i is the i-th projection. Then U is a finitely generated faithful uniform module.

The converse follows because uniform modules are primary. ■

Corollary 2.5. Every right FBN ring R has a primary decomposition such that each primary component is an ideal.

Proof. Let $B_1 \cap \ldots \cap B_n = 0$ be a primary decomposition of R. Then R/B_i and $R/\text{ann}\,(R/B_i)$ have the same associated prime and so $\text{ann}\,R/B_1 \cap \ldots \cap \text{ann}\,R/B_n = 0$ is also a primary decomposition. ■

We have not been able to prove an analogous result for finitely generated modules M. The first obstruction is that if $\cap M_i = 0$ is a primary decomposition of M and $T_i = \text{ann}\,M/M_i$ then M/MT_i need not be primary; but see Lemma 3.2. However if M is faithful then $\cap T_i = 0$ and so, as promised, we do get

Corollary 2.6. If M is a finitely generated faithful module over a right FBN ring R then ass R $\subseteq$ ass M.

Using these results we have, in [4, Corollary 2.6 (ii)], obtained some useful criteria for telling when a prime ideal is an associated prime of a given module.

Corollary 2.7. If M is a finitely generated R-module, R a right FBN ring, then the following are equivalent properties of a prime ideal P of R.

(i) $P \in$ ass M.

(ii) R/P embeds in a finite direct sum of copies of M.

(iii) $\text{ann}_R\,(\text{ann}_M P) = P$.

<u>Corollary 2.8</u>. A prime right FBN ring embeds in a finite direct sum of copies of any faithful finitely generated module.

We end the section with an example of a primary right FBN ring whose associated prime is embedded. We use a factor ring, R say, of a ring constructed by Jategaonkar (see [4, Example 4.1]). The lattice of right ideals of R is linearly ordered and looks like

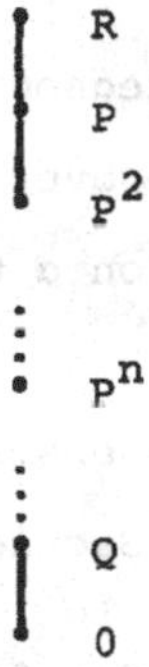

Every right ideal is principal and two-sided. We see from the picture that P is the unique maximal right ideal of R, and thus that $Q \simeq R/P$. Hence R is a P-primary right FBN ring. Furthermore the picture makes it plain that P is not isolated.

We would like to be able to show that if the associated prime of a finitely generated primary module over a right FBN ring is isolated, then the module is Macaulay. However we can do so only in special cases.

§3. FBN RINGS

Our first result shows that the pathology exhibited by the example at the end of the preceding section cannot occur when the underlying ring is an FBN ring.

Theorem 3.1. Every primary module over an FBN ring is Macaulay. In particular the associated prime of a finitely generated primary module is isolated.

This is a consequence of Jategaonkar's key result [7] which says that if a finitely generated module over an FBN ring has a critical composition factor of Krull dimension α then it has an α-critical submodule.

Next we prove a result which ensures the existence of invariant primary components, as promised in Section 1.

Lemma 3.2. A P-primary submodule of a finitely generated module M over an FBN ring contains a largest invariant P-primary submodule of M.

Proof. Let N be a P-primary submodule of M and let M_0 be the largest invariant submodule of M contained in N. Note that $T_P(M/M_0)$ is an invariant submodule of M/M_0 contained in N/M_0. But each endomorphism of M induces an endomorphism of M/M_0. It follows that $T_P(M/M_0) = 0$.

Let I = ann M/N. Then, by the choice of M_0, I is the annihilator of M/M_0 too. Thus M/M_0 cannot have a critical submodule of Krull dimension > K dim R/I, where R is the ring in question. But K dim R/I = K dim M/N = K dim R/P, by 3.1, and so M/M_0 is P-primary. ∎

Corollary 3.3. If P is not a minimal element of spec M, M a finitely generated module over an FBN ring, then every P-primary submodule of M contains an infinite descending chain of invariant P-primary submodules of M.

Proof. Let N be a P-primary submodule of M. By 1.2 M contains an infinite descending chain of P-primary submodules and, using 3.1, M/N contains no such infinite descending chain. Thus N must contain an infinite descending chain $N_1 \supset N_2 \supset \ldots$ of P-primary submodules N_i.

Let A_1 be the largest invariant P-primary submodule of M contained in N_1. By the above argument, $A_1 \not\subseteq \cap N_i$. Thus there is a first integer, i_1 say, such that $A_1 \not\subseteq N_{i_1}$. Let A_2 be the largest invariant P-primary submodule of M contained in N_{i_1}. Of course $A_2 \subset A_1$. Next choose i_2 and $A_3 \subsetneq N_{i_2}$ such that $A_3 \subset A_2$; and so on. ∎

Corollary 3.4. A primary decomposition of a finitely generated module M over an FBN ring can be chosen so that each primary component is an invariant submodule of M.

Corollary 3.5. Every FBN ring is a subring of an Artinian ring.

Using 3.4 (or 2.5) it will suffice to show that a primary FBN ring is an order in an Artinian ring. We will prove some more general results.

Let R be a right Noetherian ring with prime radical N = N(R). We say that a module M is N-torsion (respt. N-torsionfree) if it torsion (respt. torsionfree) with respect to the torsion theory cogenerated by R/N; and we denote the largest N-torsion submodule of M by $T_N(M)$. We call a nonzero module N-critical if it is N-torsionfree and every proper factor module is N-torsion. Finally we recall that a compressible module is a nonzero module which embeds in each of its nonzero submodules.

Proposition 3.6. Let R be a Noetherian ring of Krull dimension α such that R/N(R) is Macaulay.

(i) A finitely generated module M is N-torsion if and only K dim M < α.

(ii) $R/T_N(R)$ is an order in an Artinian ring if and only if every finitely generated α-critical module is compressible.

<u>Proof</u>. (i) If M is not N-torsion then some subfactor module of M has the same Krull dimension as a nonzero submodule of R/N. Therefore, since R/N is Macaulay, $K \dim M \geq K \dim R/N = \alpha$.

Conversely suppose that M is N-torsion and that $K \dim M \not< \alpha$. Then M has an α-critical N-torsion subfactor module C which is killed by N (since N is nilpotent). But then some nonzero submodule of C embeds in the N-torsionfree module R/N.

(ii) Clearly $T_N(R)$ is contained in N and is the largest right ideal of R each element of which is killed by an element of R which is regular modulo N. Thus by Small's Theorem, or directly, $R/T_N(R)$ is an order in an Artinian ring if and only if R is localizable at N. Moreover, by Jategaonkar's localizability criterion [8, Theorem 3.3], R is localizable at N if and only if every finitely generated N-critical module is annihilated by N.

By (i), a finitely generated module is N-critical precisely when it is α-critical. Thus it remains to show that a finitely generated α-critical module is compressible precisely when it is killed by N. But one implication is clear because N is nilpotent and the other because a finitely generated α-critical module killed by N is isomorphic to a right ideal of R/N. ∎

Since finitely generated critical modules over FBN rings are compressible [7], we have

<u>Corollary 3.7</u>. If R is an FBN ring such that R/N is Macaulay then R is localizable at N (or, put otherwise, $R/T_N(R)$ is an order in an Artinian ring).

Finally we have

Theorem 3.8. An FBN Macaulay ring is an order in an Artinian ring.

For this it suffices to prove that if an FBN ring R is Macaulay then so too is R/N. But since the minimal elements of spec R are precisely the equivalence classes of critical right ideals of R/N (because spec R = spec R/N), this is an immediate consequence of

Proposition 3.9. If M is a finitely generated module over an FBN ring and C a minimal element of spec M then C ε comp M. Thus K dim C = K dim D for some D ε ass M.

Proof. Since M is not C-torsion some critical composition factor of M is not C-torsion. The compressibility of this critical composition factor then forces it to be C-torsionfree. ■

Of course a minimal prime ideal of an FBN ring need not be an associated prime ideal of the ring.

We recall that a module is said to be torsionfree if no nonzero element of the module is killed by a regular element of the ring.

Corollary 3.10. Let R be an FBN ring and M a finitely generated nonzero module.

(i) If R is Macaulay then M is torsionfree if and only if M is Macaulay and K dim M = K dim R.

(ii) If M is Macaulay then M is torsionfree if and only if every regular element of R is regular modulo ann M.

Proof. (i) Since R is an order in an Artinian ring, Small's Theorem implies that M is torsionfree precisely when it is N(R)-torsionfree. Thus the result is a consequence of 3.6 (i).

(ii) Let A = ann M. Then, of course, R/A is Macaulay. Thus

every left regular element of R/A is regular. But if M is torsionfree then every regular element of R is left regular modulo A.

The converse is clear because M is torsionfree over R/A, by (i). ■

We should mention T. Lenagan has been able to prove Theorem 3.8 without using the fact that finitely generated critical modules are compressible, and that this leads immediately to a new proof of this crucial result. Moreover he has shown that any Noetherian Macaulay ring of Krull dimension 1 is an order in an Artinian ring.

We should also mention the basic reason for our interest in Macaulay modules and rings. This is that any nonzero Noetherian module M determines uniquely a sequence $0 \neq M_1 \subset M_2 \subset \ldots \subset M_n = M$ of submodules of M and a sequence $-1 \neq \alpha_1 < \alpha_2 < \ldots < \alpha_n$ of ordinals such that M_i/M_{i-1} is α_i-Macaulay for each i ($M_0 = 0$). We refer to the former sequence as the <u>submodule sequence</u> of M and to the latter as the <u>Krull dimension sequence</u> of M. Note that each M_i is an invariant submodule of M. In fact the submodule sequence can be defined inductively by setting $M_i/M_{i-1} = \tau(M/M_{i-1})$. The first part of the next result shows, that for a finitely generated module over an FBN ring, it can be gotten from a primary decomposition of the module.

<u>Theorem 3.11</u>. Let $\cap N_j = 0$ be a primary decomposition of a finitely generated module M over an FBN ring and let $\alpha_1 < \ldots < \alpha_n$ be the Krull dimension sequence of M. If

$$M_i = \cap\{N_j \mid K \dim M/N_j > \alpha_i\},$$

$$H_i = \cap\{N_j \mid K \dim M/N_j \neq \alpha_i\},$$

then

(i) $M_1 \subset \ldots \subset M_n = M$ is the submodule sequence of M,

(ii) $\cap_j\{N_j/M_i \mid K \dim M/N_j > \alpha_i\} = 0$ is a primary decomposition of M/M_i, and

(iii) H_i is an α_i-Macaulay submodule of M_i such that $K \dim M_i/H_i < \alpha_i$.

<u>Proof</u>. We observe that since the submodule sequence of M can be refined to a critical composition series, the α_i are precisely the distinct Krull dimensions of the M/N_j.

(i) Our observation ensures, using 1.4, that $M_i/M_{i-1} \neq 0$. It also implies that

$$M_{i-1} = M_i \cap \cap\{N_j \mid K \dim M/N_j = \alpha_i\}.$$

From this it follows that M_i/M_{i-1} embeds in the α_i-Macaulay module $\coprod\{M/N_j \mid K \dim M/N_j = \alpha_i\}$.

(ii) This is a consequence of Lemma 1.4.

(iii) Note that

$$H_i = M_i \cap \cap\{N_j \mid K \dim M/N_j < \alpha_i\}. \quad (1)$$

This makes it plain that $H_i \cap M_{i-1} = 0$ and thus that H_i embeds in M_i/M_{i-1}. Hence H_i is α_i-Macaulay, by (i).

It remains to prove that $K \dim M_i/H_i < \alpha_i$. Of course $H_1 = M_1$. But then, if $i > 1$, the formula (1) shows that M_i/H_i is isomorphic to a submodule of $\coprod\{M/N_j \mid K \dim M/N_j < \alpha_i\}$--and this has Krull dimension $< \alpha_i$. ∎

We point out that, roughly speaking, (iii) says that the terms of the submodule sequence can be approximated (in the sense of Krull dimension) by Macaulay modules. We also point out that the H_i may be chosen to be invariant submodules of M, by Corollary 3.4.

We claim that comp M = $\cup$comp H_i. For $H_i \simeq \overline{H}_i \subseteq \overline{M}_i = M_i/M_{i-1}$ and $K \dim \overline{M}_i/\overline{H}_i < K \dim \overline{M}_i$. Thus comp H_i = comp M_i by the proof of Lemma 1.9. (Alternately, note that $\overline{M}_i$ and $\overline{H}_i$ have the same annihilator and use 1.10.)

Similarly spec M = $\cup$ spec H_i and ass M = $\cup$ ass H_i.

Lastly, note that $\bar{H}_i$ is essential in $\bar{M}_i$. Thus, given any torsion theory T, M is T-torsionfree precisely when every H_i is T-torsionfree.

The proof of the following result illustrates the usefulness of these facts.

<u>Corollary 3.12</u>. If R is an RBN order in an Artinian ring and M a finitely generated module then M is torsionfree if and only if comp M $\subseteq$ comp R. Moreover, in this case comp M is the set of minimal elements of spec M.

<u>Proof</u>. We may assume, for the proof of the first statement, that M is Macaulay.

=> Let C ε comp M. Then C is torsionfree over R/ann M, by 3.10 (i), and hence torsionfree over R by 3.10 (ii). Thus C is N(R)-torsionfree. But R/N and $\coprod \{D \mid D$ a minimal element of spec R$\}$ cogenerate the same torsion theory. It follows that C is a minimal element of spec R. Thus C ε comp R by 3.9.

<= Since R itself is torsionfree this follows from the proof just given.

The proof of the second statement is a straightforward consequence of the above analysis, using Proposition 3.9. ∎

The last three results are essentially in [5], and have many consequences in that paper.

REFERENCES

1. P. Gabriel, Des catégoies abéliennes, Bull. Soc. Math. France 90(1962), 323-448.

2. R. Gordon, Gabriel and Krull dimension, in "Ring Theory" (B. R. McDonald, A. R. Magid, K. C. Smith, Eds.), Lecture Notes in Pure and Applied Math. 7, Marcel Dekker, New York (1974), 241-295.

3. R. Gordon, Artinian quotient rings of FBN rings, J. Algebra 35(1975), 304-307.

4. R. Gordon, Primary decomposition in right noetherian rings, Communications in Algebra 2(1974), 491-524.

5. R. Gordon and E. L. Green, A representation theory for noetherian rings, J. Algebra, to appear.

6. R. Gordon and J. C. Robson, Krull dimension, Memoirs Amer. Math. Soc. 133(1974).

7. A. V. Jategaonkar, Jacobson's conjecture and modules over fully bounded noetherian rings, J. Algebra 30(1974), 103-121.

8. A. V. Jategaonkar, Injective modules and localization in non-commutative noetherian rings, Trans. Amer. Math. Soc. 188(1974), 109-123.

CERTAIN INJECTIVES ARE ARTINIAN

Arun Vinayak Jategaonkar
Fordham University
Bronx, New York

The main result of this paper shows that the injective hull of a simple module over a Noetherian P.I. ring is an Artinian module with finitely many chief factors. This refines a portion of our earlier result [2] and includes a result of Rosenberg and Zelinsky [7] concerning injectives over Artinian P.I. rings. Our main result implies that a Noetherian P.I. ring has a right Morita duality if and only if it is a complete semi-local ring.

We also show that if S is a simple right module over a right Noetherian ring R such that $\text{ann}_R S$ has a normalizing set of generators and $R/\text{ann}_R S$ is an Artinian ring then the R-injective hull $E_R(S)$ of S is an Artinian module. It then follows (cf. [1]) that $E_R(S)$ is the union of its socle sequence. This information may be of interest in connection with enveloping algebras of solvable Lie algebras and group algebras of f.g. nilpotent groups (cf. [3, 6]).

Our proof of the main result depends upon a result of Rowen [8] and our earlier results about modules over Noetherian P.I. rings [2]. We also need an apparently new technical device which we proceed to explain.

Our technical device compares a given module E with the graded module associated with a filtration on E which is dual to the usual I-adic filtration. Thus, let I be a two-sided ideal of a ring R and E be a right R-module. For each non-negative integer n, we define $E_n = \text{ann}_E I^n$. We can then define a structure of a graded right $\text{gr}_I R$-module on $\bigoplus_{n=1}^{\infty}(E_n/E_{n-1})$ by the following rule: If $\chi_n \in E_n$ and $r_m \in I^m$ then

$$[\chi_n + E_{n-1}][r_m + I^{m+1}] = \begin{cases} [\chi_n r_m + E_{n-m-1}] & \text{if } n \geq m+1 \\ 0 \text{ otherwise} \end{cases}$$

We shall denote this module as $\mathrm{cogr}_I E$ (or cogr E) to distinguish it from the usual graded module associated with the I-adic filtration on E. For each R-submodule F of E we can and shall identify cogr F with the homogeneous grR-submodule

$$\bigoplus_{n=1}^{\infty} [(F + E_{n-1}) \cap E_n]/E_{n-1}$$

of cogr E. Thus the map F -> cogr F defines an increasing function from the partially ordered set of R-submodules of E to the partially ordered set of homogeneous grR-submodules of cogr E. The situation of particular interest is the one in which this function is strictly increasing. It is characterized in the following lemma.

At this point, we need a definition. By a <u>chief factor</u> of a module M we mean a simple image of a submodule of M. A module M is said to have <u>finitely many chief factors</u> if the set of mutually non-isomorphic chief factors of M is finite.

Suppose M is the union of an ascending chain of submodules $\{M_n : n \geq 0\}$ with $M_o = (0)$. Let S be a chief factor of M. Since S can be realized as an image of a cyclic submodule of M, it is also a chief factor of some M_n. Choose n minimally and choose submodules $D \subset cR = C \subseteq M_n$ such that $C/D \cong S$. If $C \cap M_{n-1} \nsubseteq D$ then $C/D \cong (C \cap M_{n-1})/(D \cap M_{n-1})$, contrary to our choice of n. Thus $C \cap M_{n-1} \subseteq D$ and S is a chief factor of M_n/M_{n-1}. It is now easy to see that M and $\bigoplus_{n=1}^{\infty} (M_n/M_{n-1})$ have the same chief factors. We shall use this fact without further mention.

<u>Lemma 1</u>. Let I be a two-sided ideal of a ring R and E be a right R-module. Then

(a) The function sending an R-submodule F of E to the $gr_I R$-submodule $cogr_I F$ of $cogr_I E$ is strictly increasing if and only if $E = \cup\{ann_E I^n : n \geq 1\}$. If these equivalent conditions are satisfied then E has finitely many chief factors if and only if $cogr_I E$ has finitely many chief factors.

(b) $ann_E I$ is an essential $gr_I R$-submodule of $cogr_I E$.

<u>Proof</u>. For each R-submodule F of E, we set $F_n = ann_F I^n$. Let $F \subseteq G$ be R-submodules of E such that cogr F = cogr G. Inductively, assume that $F_n = G_n$. Since

$$\begin{aligned} F_{n+1} + E_n &= (F + E_n) \cap E_{n+1} \\ &= (G + E_n) \cap E_{n+1} \\ &= G_{n+1} + E_n \end{aligned}$$

and

$$G_{n+1} \cap E_n = G_n = F_n,$$

it follows that $F_{n+1} = G_{n+1}$. Thus, if $E = \bigcup_{n=1}^{\infty} E_n$ then $F = \bigcup_{n=1}^{\infty} F_n = \bigcup_{n=1}^{\infty} G_n = G$. It is now easy to complete the proof of part (a).

For (b), we observe that any non-zero element χ of cogr E can be expressed as $\chi_{n_1} + \ldots + \chi_{n_m}$ where $n_1 < \ldots < n_m$ and each χ_{n_i} is a non-zero element of $E_{n_i}/E_{n_i - 1}$. Evidently, if m = 1 then $\chi \in E_1$ and if m > 1 then

$$0 \neq \chi(I^{n_m - 1}/I^{n_m}) = \chi_{n_m}(I^{n_m - 1}/I^{n_m}) \subseteq E_1.$$

Thus E_1 is essential in cogr E. ■

We recall that an element χ of a ring R is called a <u>normalizing</u>

element of R if $R\chi = \chi R$. A finite indexed subset $\{\chi_1, \ldots, \chi_n\}$ of R is called a normalizing set (resp. centralizing set) in R if the image of each χ_i in $R/\sum_{j=1}^{i-1} R\chi_j R$ is a normalizing element (resp. a central element) of $R/\sum_{j=1}^{i-1} R\chi_j R$.

Lemma 2. Let χ be a normalizing element in a right Noetherian ring R and E be a right R-module such that $\mathrm{ann}_E(\chi R)$ is Artinian and essential in E. Then E is an Artinian R-module and each chief factor of E_R is Z(R)-isomorphic with a chief factor of the R-module $\mathrm{ann}_E(\chi R)$. Moreover, if χ is central in R then E and $\mathrm{ann}_E(\chi R)$ have the same chief factors.

Proof. As before, we let $E_n = \mathrm{ann}_E(\chi^n R)$. Let $H_n = E_n/E_{n-1}$ and let $f_n : H_{n+1} \to H_n$ be the map induced by right multiplication by the element χ. We observe that f_n is a monomorphism of Z(R)-modules, Z(R) being the center of R. If $\chi \in Z(R)$ then f_n is R-linear too. In general, f_n need not be R-linear but it still sends R-submodules to R-submodules.

Now let $\{F_n : n \geq 1\}$ be a descending chain of R-submodules of E. Then $\{\mathrm{cogr}_{\chi R} F_n : n \geq 1\}$ is a descending chain of homogeneous grR-submodules of cogr E. Express each cogr F_n as $\bigoplus_{m=1}^{\infty} K_{nm}$ where each $K_{nm} \subseteq H_m$ and define

$$L_{nm} = \begin{cases} K_{nm} & \text{if } m = 1, \\ f_1 \ldots f_{m-1}(K_{nm}) & \text{if } m > 1. \end{cases}$$

It is clear that each L_{nm} is an R-submodule of $\mathrm{ann}_E(\chi R)$ and that L_{nm} decreases if n or m is increased. Since $\mathrm{ann}_E(\chi R)$ is Artinian, the descending chain $\{L_{nn} : n \geq 1\}$ eventually stops. It is now easy to see that the descending chain $\{\mathrm{cogr}\ F_n : n \geq 1\}$ eventually stops. Since it is known (cf. Proof of Lemma 8 of [4]) that $E = \bigcup_{n=1}^{\infty} E_n$, Lemma 1 now shows that the descending chain $\{F_n : n \geq 1\}$ eventually stops. The assertions about chief factors follow immediately from the properties of f_n observed above. ■

Theorem 1. Let I be an ideal of a right Noetherian ring R and let S be a simple right R-module which is annihilated by I. If I has a normalizing set of generators then the R-injective hull $E_R(S)$ of S is Artinian if and only if the R/I-injective hull $E_{R/I}(S)$ of S is Artinian. In this situation, each chief factor of $E_R(S)$ is Z(R)-isomorphic with a chief factor of $E_{R/I}(S)$. Moreover, if I has a centralizing set of generators then $E_R(S)$ and $E_{R/I}(S)$ have the same chief factors.

Proof. Follows immediately from Lemma 2. ■

The result stated in the second paragraph of the introduction is obviously a special case of Theorem 1.

We now turn to Noetherian P.I. rings. Our proof of the main result utilizes a Noetherian induction. We now have enough tools to complete the induction if the ring under consideration happens to be semi-prime. The information needed to handle the non-semi-prime case is given by Lemma 5 below. Lemmas 3 and 4 are needed to prove Lemma 5.

The following lemma is essentially contained in [7].

Lemma 3. Let R and S be Artinian P.I. rings, ${}_SX_R$ be an (S,R)-bimodule which is finitely generated from left as well as right and E be a right R-module with finitely generated socle. Then the right S-module $\mathrm{Hom}({}_SX_R, E_R)$ is finitely generated.

Proof. Firstly, we shall consider the case in which X is semi-simple and homogeneous from left as well as right. We may then assume without loss of generality that ${}_SX$, X_R and E_R are faithful modules. This makes R a simple Artinian P.I. ring. Using Kaplansky's Theorem, it follows that the center F of R is a field and that R_F, X_F and E_F are finite dimensional spaces over F. Since $\mathrm{Hom}({}_SX_R, E_R)$ is an S-submodule of $\mathrm{Hom}({}_SX_F, E_F)$ and $\mathrm{Hom}\ ({}_SX_F, E_F)$ is a finite direct sum of $\mathrm{Hom}({}_SX_F, F_F)$ with itself, it suffices to show that $\mathrm{Hom}({}_S\ {}_F, F_F)$ is finitely generated as a right S-module.

Let $k = \dim X_F$. We shall realize X_F as the space $M_{k\times 1}(F)$ of $k \times 1$ matrices over F, $\mathrm{Hom}\ (X_F, F_F)$ as the space $M_{1\times k}(F)$ and End X_F as the ring $M_k(F)$. Since ${}_SX$ is faithful, the left action of S on X induces an injective ring homomorphism $\phi: S \to M_k(F)$. We shall identify S with $\phi(S)$ by ϕ. Then the right action of S on $\mathrm{Hom}({}_SX_F, F_F)$ becomes the action of S on $M_{1\times k}(F)$ derived from the action of $M_k(F)$. Since ${}_SX$ is f. g., the first column of $M_k(F)$ is f. g. as a left S-module. This makes $M_k(F)$ a f.g. left S-module. Consequently, $F\cdot S$ is f.g. as a left S-module. So, there exist $z_1, \ldots, z_n \in F$ such that

$$F\cdot S = \sum_{i=1}^{n} Sz_i = \sum_{i=1}^{n} z_i S.$$

Thus $F\cdot S$ is f.g. as a right S-module. Since $M_k(F)$ is evidently a f.g. right $F\cdot S$-module, it follows that $M_k(F)$ is a f.g. right S-module. So, $M_{1\times k}(F)$ is f.g. as a right S-module. This completes the proof of the special case under consideration.

We now turn to the general case. Let W be a maximal bisubmodule of X. Then X/W is semi-simple and homogeneous from left as well as right. Using an induction on the length of X_R and the exact sequence

$$0 \to \mathrm{Hom}({}_S(X/W)_R, E_R) \to \mathrm{Hom}({}_SX_R, E_R) \to \mathrm{Hom}({}_SW_R, E_R)$$

it is easy to complete the proof. ■

<u>Lemma 4</u>. Let R be a Noetherian P.I. ring, X be an R-bimodule which is finitely generated from left as well as right and E be an injective Artinian right R-module with finitely many chief factors. Then the socle of the right R-module $\mathrm{Hom}({}_RX_R, E_R)$ is finitely generated.

Proof. Let $A_1, \ldots, A_k$ be the annihilators of representatives of isomorphism classes of chief factors of E. By Kaplansky's Theorem, each R/A_i is a simple Artinian ring; so, the R-bimodule X/XA_i is f.g. and semi-simple as a right module. Let

$$(0) = Y_{io} \subset Y_{i1} \subset \cdots \subset Y_{in_i} = X/XA_i$$

be a bimodule composition series of X/XA_i. Set $X_{ij} = Y_{ij}/Y_{ij-1}$, $1 \leq j \leq n_i$. By Jordan-Holder Theorem, each simple subfactor bimodule of X/XA_i is bimodule isomorphic with some X_{ij}. Lemma 2.2 of [2] implies that X/XA_i is of finite length as a left R-module. So, each X_{ij} has to be semi-simple and homogeneous as a left R-module. Using Kaplansky's Theorem, it follows that the left annihilator B_{ij} of X_{ij} is a maximal ideal of R. Let B be the intersection of all B_{ij} as i and j vary. Evidently R/B is a semi-simple ring.

Now let $f \cdot R$ be a simple submodule of $\mathrm{Hom}({}_RX_R, E_R)$. The main step in the proof of the lemma is to show that $BX \subseteq \ker f$. We need a few observations first.

Let W be the largest bisubmodule of X contained in ker f. Let $\bar{X} = X/W$ and $\bar{f}:\bar{X} \to E$ be the homomorphism induced by f. Clearly, $\ker \bar{f}$ does not contain any non-zero bisubmodule of $\bar{X}$. Since $\bar{f}(\bar{X})$ is a f.g. submodule of E, it is of finite length. It follows that, for some positive integer m, the bisubmodule $\bar{X}(\bigcap_{i=1}^{k} A_i)^m$ of $\bar{X}$ is contained in $\ker \bar{f}$. So $\bar{X}(\bigcap_{i=1}^{k} A_i)^m = (0)$. It follows that $\bar{X}_R$ is of finite length and that each chief factor of $\bar{X}_R$ is annihilated by some A_i.

We now look at the left R-module $\bar{X}$. Let C be its annihilator in R. Since $\bar{X}_R$ is of finite length, Lemmas 2.1 and 2.2 of [2] show that ${}_R\bar{X}$ is of finite length and R/C is an Artinian ring. Assume for a moment that $\bar{f} \cdot J(R/C) \neq (0)$; say $\bar{f} \cdot [r + C] \neq 0$ for some $[r + C] \in J(R/C)$. Then $f \cdot r \neq 0$; so the assumed simplicity of $f \cdot R$ provides us an element $r' \in R$ such that $f = frr'$. Thus

$$\bar{f}\cdot([1 + C] - [r + C][r' + C]) = 0.$$

This yields $\bar{f} = 0$ and so $f = 0$, contrary to the assumed simplicity of $f\cdot R$. Hence $\bar{f}\cdot J(R/C) = (0)$; i.e., $J(R/C)\bar{X} \subseteq \ker \bar{f}$. Since $\ker \bar{f}$ does not contain any non-zero bisubmodule of $\bar{X}$, we have $J(R/C)\bar{X} = (0)$ and so $J(R/C) = 0$. Hence R/C is a semi-simple ring and ${}_R\bar{X}$ is a f.g. semi-simple module.

We now show that $BX \subseteq \ker f$ by showing that $B \subseteq C$. Let $e_1, \ldots, e_t$ be the central indecomposible idempotents of the semi-simple ring R/C. For $1 \leq n \leq t$, let $V_n = e_n\bar{X}$ and let C_n be the annihilator in R of the left R-module V_n. It is clear that each V_n is an R-bimodule which is semi-simple and homogeneous as a left R-module and that the sum $\bar{X} = \sum_{n=1}^{t} V_n$ is a direct sum of R-bimodules. It is also clear that $C = \bigcap_{n=1}^{t} C_n$. Now, let V_n' be a maximal bisubmodule of V_n. Then $V_n/V_n' = \bar{V}_n$ is a simple factor bimodule of $\bar{X}$. So, it has to be semi-simple and homogeneous as a right R-module. But, as shown above, every chief factor of $\bar{X}_R$ is annihilated by some A_i. Thus, $\bar{V}_nA_i = (0)$ for some i. It is now clear that $\bar{V}_n$ is bimodule isomorphic with some X_{ij}. So, $B_{ij}\bar{V}_n = (0)$. The homogeneity of the left R-module V_n now yields $B_{ij}V_n = (0)$; so $B_{ij} \subseteq C_n$. Hence $B \subseteq C$ and $BX \subseteq \ker f$.

Since the bisubmodule BX does not depend upon f, it follows that the socle of the right R-module $\mathrm{Hom}({}_RX_R, E_R)$ is isomorphic with the socle of the right R-module $\mathrm{Hom}({}_R(X/BX)_R, E_R)$. Let I be the annihilator of the right R-module X/BX. Since R/B is semi-simple, Theorem 2.3 of [2] implies that R/I is an Artinian ring. Lemma 3 now shows that the right R-module $\mathrm{Hom}({}_R(X/BX)_R, \mathrm{ann}_E I)$ is of finite length. ∎

<u>Lemma 5</u>. Let R be a Noetherian P.I. ring such that the injective hull of each simple R-module is an Artinian module with finitely many chief factors. Let X be an R-bimodule which is finitely generated from left

as well as right. Then the split extension T of R by X is a Noetherian P.I. ring and the injective hull of each simple T-module is an Artinian module with finitely many chief factors.

Proof. Recall that the additive group of T is $R \oplus X$ and the multiplication is defined by the rule

$$(r,\chi)(r',\chi') = (rr', r\chi' + \chi r')$$

for all $r, r' \in R$ and $\chi, \chi' \in X$. It is thus obvious that T is a Noetherian P.I. ring. Let E be the R-injective hull of a simple right R-module. To complete the proof, it suffices to show that the T-injective hull of E_T is an Artinian module with finitely many chief factors.

Consider the right T-module $H = E \oplus \mathrm{Hom}({}_RX_R, E_R)$ where the action of T is given by the rule

$$(e, f)(r, \chi) = (er + f(\chi), fr).$$

It is straightforward to check that the obvious map

$$\mathrm{Hom}({}_TT_R, E_R) \to E \oplus \mathrm{Hom}({}_RX_R, E_R)$$

is an isomorphism of right T-modules. Using the injectivity of E_R, it follows (cf. p. 131 of [3]) that H_T is an injective T-module. Evidently, the T-submodule (E, 0) of H is T-isomorphic with E treated naturally as a T-module. Let I be the T-injective hull of (E, 0) in H. Then I is also the T-injective hull of a simple T-module. So, by Corollary 3.6 of [2], finitely generated submodules of I_T are of finite length. Notice that $I = E \oplus F$ where

$$F = \{f \in \mathrm{Hom}({}_RX_R, E_R) \mid (0, f) \in I\}.$$

Clearly, F is an R-submodule of $\mathrm{Hom}({}_RX_R, E_R)$ and is isomorphic with I/(E, 0) treated as an R-module. It follows that finitely generated submodules of F_R are of finite length. In particular, F_R is an essential extension of its socle. Using Lemma 4, it is immediate that soc F_R is finitely generated. So, F_R is an Artinian module with finitely many chief factors. Since $I = E \oplus F$, it follows that I_T is an Artinian module with finitely many chief factors. ■

We now prove our main result.

<u>Theorem 2</u>. The injective hull of a simple module over a Noetherian P.I. ring is an Artinian module with finitely many chief factors.

<u>Proof.</u> Let R be a Noetherian P.I. ring. Due to Noetherian induction, we may assume that the theorem holds for every proper factor ring of R. To show that the theorem then holds for R, we take the injective hull E of a simple right R-module S and proceed to show that E is Artinian with finitely many chief factors.

Suppose R is a semi-prime ring. If $\mathrm{ann}_R S = (0)$ then, by Kaplansky's Theorem, R has to be simple Artinian. The conclusion we want is trivial in this case. So, we may as well assume that $\mathrm{ann}_R S \neq (0)$. Then, by a result of Rowen [8], $\mathrm{ann}_R S$ contains a non-zero central element of R, say χ. Let $F = \mathrm{ann}_E(\chi R)$. Since $S\chi = (0)$, E is an essential extension of F. Moreover, when treated naturally as a right module over $R/\chi R$, F is the injective hull of S. By our Noetherian induction hypothesis, F_R is Artinian with finitely many chief factors. So, by Lemma 2, E_R is Artinian with finitely many chief factors.

Suppose R is not a semi-prime ring. Then, using the nilpotency of the prime radical, we can choose a non-zero two-sided ideal I of R such that $I^2 = (0)$. Lemma 1(b) shows that $\mathrm{cogr}_I E$ is an essential extension of a simple right module over $\mathrm{gr}_I R$. But $\mathrm{gr}_I R$ is just the split extension of R/I by I treated naturally as an (R/I)-bimodule. Thus, using the Noetherian induction hypothesis and Lemma 5, it follows that the $gr_I R$-module $cogr_I E$ is Artinian with finitely many chief factors. Lemma 1(a) now shows that E_R is Artinian with finitely many chief factors. ■

Theorem 3. A Noetherian P.I. ring has a right Morita duality if and only if it is a complete semi-local ring.

Proof. The 'if' part follows from Theorem 8 of Muller [5] and Theorem 2 above. The 'only if' part follows from Theorem 7 of [5] and Theorem 3.7 of [2]. ■

REFERENCES

1. J. P. Jans, On co-Noetherian rings, J. London Math. Soc. (2), 1(1969), 588-590.

2. A. V. Jategaonkar, Jacobson's conjecture and modules over fully bounded Noetherian rings, J. Algebra, 30(1974), 103-121.

3. J. Lambek, Lectures on rings and modules, Blaisdell Publishing Co., Walthan, Mass. 1966.

4. J. C. McConnell, Localization in enveloping rings, J. London Math. Soc. 43(1968), 421-428; Erratum and addendum, ibid., (2), 3(1971), 409-410.

5. B. J. Muller, On Morita duality, Canad. J. Math., 6(1969), 1338-1347.

6. J. E. Roseblade, The integral group rings of hypercentral groups, Bull. London Math. Soc., 3(1971), 351-355.

7. A Rosenberg and D. Zelinsky, Finiteness of the injective hull, Math. Z., 70(1959), 372-380.

8. L. Rowen, Some results on the center of a ring with polynomial identity, Bull. Amer. Math. Soc., 79(1973), 219-223.

ARTINIAN QUOTIENT RINGS OF MACAULAY RINGS

T. H. Lenagan

Mathematical Institute

Chambers Street, Edinburgh, Scotland

In a recent paper [4], Gordon has shown that certain kinds of fully bounded Noetherian rings, which he calls Macaulay rings, have Artinian quotient rings. His proof uses the results of Jategaonkar [7], on basic composition series of fully bounded Noetherian rings. In a later paper [5], Gordon points out that some of Jategaonkar's results follow from the existence of Artinian quotient rings. Of course, he admits, the reasoning is circular, and so he appeals for a more direct proof of his earlier result. In Section 2 of this paper we provide such a proof and in Section 3 we modify the arguments in order to encompass certain rings with Krull dimension one (such rings need not be fully bounded). In Section 4 we present a short proof of the recent result of Cauchon [1], and Jategaonkar [7], that the Jacobson Conjecture holds for left Noetherian, right fully bounded Noetherian rings. We shall assume throughout the paper that rings have identity elements.

§1. Definitions and Preliminary Results

A ring is <u>right bounded</u> if every essential right ideal contains a non-zero two-sided ideal. A right Noetherian ring is <u>fully right bounded Noetherian</u> if every prime factor ring is right bounded. A ring is <u>fully bounded Noetherian</u> if it is fully right bounded Noetherian and fully left bounded Noetherian.

Let M be a right R-module. The <u>Krull dimension</u> of M denoted by $|M|$, is defined as follows: if $M = 0$ then $|M| = -1$; if α is an ordinal and $|M| \not< \alpha$, then $|M| = \alpha$ provided that there is no infinite descending chain $M = M_0 \supsetneq M_1 \supsetneq \dots$ of submodules M_i such that, for

$i = 1, 2, \ldots, |M_{i-1}/M_i| \not< \alpha$. It is possible that there is no ordinal α such that $|M| = \alpha$. In this case we say that M has no Krull dimension. The (right) Krull dimension of a ring R is defined to be $|R_R|$ and is written $|R|$.

We refer the reader to [6] for basic properties of Krull dimension, while we list, without proof, some of the results that we need.

Lemma 1.1. If M is a module with Krull dimension and N is a submodule of M then $|M| = \sup\{|N|, |M/N|\}$.

Lemma 1.2. A Noetherian module has a Krull dimension. Further, if R is a right Noetherian ring and M a finitely generated right R-module then $|M| \leq |R|$.

Corollary 1.3. If N is a nilpotent radical of a right Noetherian ring R then $|R| = |R/N|$.

Lemma 1.4. If R is a right Noetherian ring and c is a regular element of R then $|R/cR| < |R|$.

An R-module M is <u>singular</u> if, for every $m \in M$ the annihilator, $a_R(m)$, is an essential right ideal or R.

Lemma 1.5. If M is a finitely generated singular module over a right bounded, right Noetherian prime ring R, then M is unfaithful as an R-module.

Proposition 1.6 [7, Lemma 2.1]. If R is a fully right bounded Noetherian ring and M is a finitely generated R-module then $|M| = |R/\mathrm{ann}_R(M)|$.

§2. Fully Bounded Rings

The following result is a special case of [7, Lemma 2.2].

Proposition 2.1. If R is a fully bounded Noetherian ring and I a two-sided ideal of R then $|I_R| = |{}_RI|$.

In particular, for a fully bounded Noetherian ring R, we see that $|{}_RR| = |R_R|$.

If R is a Noetherian ring and β an ordinal number, we define $\tau_\beta(R)$ to be the unique largest right ideal with Krull dimension at most β. It is easy to see that $\tau_\beta(R)$ is an ideal of R. Using the previous result, we see that in a fully bounded Noetherian ring $\tau_\beta(R)$ is the largest ideal with right or left Krull dimension at most β.

Lemma 2.2. If P is a minimal prime ideal of a semiprime Noetherian ring R and $|{}^R/P| \leq \beta$ then $\tau_\beta(R)$ is non-zero.

Proof. In a semiprime Noetherian ring minimal prime ideals are annihilator ideals [3, 2.12], and so $\ell(P)$ is non-zero. But $\ell(P)$ is a right R/P-module and so $|\ell(P)| \leq |{}^R/P| \leq \beta$. ■

Lemma 2.3. If R is a right Noetherian ring and I a non-zero right ideal of R such that $\tau_\beta(R/r(I))$ is non-zero then $\tau_\beta(R)$ is non-zero.

Proof. Let K be the inverse image in R of $\tau_\beta({}^R/r(I))$ and choose $i \in I$ such that $iK \neq 0$. There is an epimorphism from K/r(I) onto iK and so $|iK| \leq |K/r(I)| \leq \beta$. ■

If A, B are subsets of a ring R we write $A.^{\cdot}B = \{r \in R \mid rB \subseteq A\}$ and $A^{\cdot}.B = \{r \in R \mid Ar \subseteq B\}$.

<u>Theorem 2.4</u>. If R is a fully bounded Noetherian ring and P is a minimal prime ideal of R such that $|R/P| \leq \beta$ then $\tau_\beta(R)$ is non-zero.

Proof. If R is a semiprime ring the result is true by Lemma 2.2. Otherwise we may as well assume that the result holds in proper factor rings of R and that N, the nilpotent radical of R, is non-zero. Put $S = P^{\cdot}.N$. Then, since P/N is a minimal prime ideal of R/N, we know, by [3, 2.12] that $S = N.^{\cdot}P$ also. As in Lemma 2.2., $S/N \subseteq \tau_\beta(R/N)$. Now $PSr(N) = 0$, so that $Sr(N)$ is a left R/P-module. Thus $S/N \subseteq \tau_\beta(R)$ and so if $Sr(N) \neq 0$ the result is proven.

If $Sr(N) = 0$, then, in particular, $Sr(N) \subseteq P$. Now $S \not\subseteq P$ and so $r(N) \subseteq P$. In this case $R/r(N)$ satisfies the conditions of the theorem and therefore, since $R/r(N)$ is a proper factor ring of R, we know that $\tau_\beta(R/r(N))$ is non-zero. Hence, by Lemma 2.3, we see that $\tau_\beta(R)$ is non-zero. ∎

Corollary 2.5. If R is a fully bounded Noetherian ring and $\tau_\beta(R/N)$ is non-zero then $\tau_\beta(R)$ is non-zero.

Proof. If $N = 0$ there is nothing to prove. Otherwise choose an ideal P maximal among annihilators of non-zero submodules of the right R-module $\tau_\beta(R/N)$. It is easy to check that P is a prime ideal and, by Proposition 1.6, we know that $|R/P| \leq \beta$. But P/N is an annihilator prime ideal in the semiprime Noetherian ring R/N and so is a minimal prime ideal of R/N and hence P is a minimal prime ideal of R. The result now follows from Theorem 2.4. ∎

The proof of the final theorem of the section is that given by Gordon [4]; we present the full proof here since we need the method in the next section.

Theorem 2.6. If R is a fully bounded Noetherian ring such that $|I| = |R|$ for each non-zero right ideal I of R then R has a two-sided Artinian quotient ring.

Proof. Any semiprime Noetherian ring has a two-sided Artinian quotient ring, and so we may as well assume that N, the nilpotent radical of R, is non-zero. Using the Noetherian condition we may assume that any proper factor ring of R that satisfies the conditions of the theorem has a two-sided Artinian quotient ring. Suppose that β is any ordinary number such that $\beta < |R|$. Then $\tau_\beta(R) = 0$ and so, by Corollary 2.5 and the left hand version of Lemma 2.3, we know that $\tau_\beta(R/N)$ and $\tau_\beta(R/\ell(N))$ are both zero. Using the natural embedding of $R/(N \cap \ell(N))$ in $R/N \oplus R/\ell(N)$, we see that $\tau_\beta(R/(N \cap \ell(N)))$ is zero. However, $N \cap \ell(N) \neq 0$ and so $R/(N \cap \ell(N))$ is a proper factor ring of R. Hence, by assumption, $R/N \cap \ell(N))$ has a two-sided Artinian quotient ring.

Now, by Small's Theorem [3, 2.7], a Noetherian ring has an Artinian quotient ring if and only if elements that are regular modulo the nilpotent radical are in fact regular elements.

Let c be an element that is regular modulo N. Then c is regular modulo $N \cap \ell(N)$, and so $\ell(c) \subseteq \ell(N)$. Suppose that $\ell(c) \neq 0$ and choose $0 \neq x \in \ell(c)$. There is an epimorphism from $R/(cR + N)$ onto xR. Now, by Corollary 1.3 and Lemma 1.4, $|R/(cR + N)| < |R/N| = |R|$. Hence we deduce that $|xR| < |R|$, which is impossible. Therefore $\ell(c) = 0$. A symmetrical argument shows that $r(c) = 0$ and so c is a regular element. ■

3. Rings With Krull Dimension One

The main difficulty in extending Theorem 2.6 to rings which are not fully bounded Noetherian rings is seen by analyzing the proof of Theorem 2.4. In this proof we isolate two possibilities: one produces a non-zero left ideal with "small" Krull dimension; the other produces a non-zero right ideal with "small" Krull dimension. We then need a result such as Theorem 2.1 to reach the desired conclusion. Thus we need some two-sided conditions; indeed, Gordon [4], gives an example of a right fully bounded Noetherian ring R with right Krull dimension one and $\tau_0(R) = 0$ which fails to have a right Artinian right quotient ring. Also we need a result such as Theorem 2.1. We have not been able to do this in general; however, we have solved the problem for the case of a right and left Noetherian ring with right Krull dimension one.

The fact that we can make progress in this low dimensional situation is a consequence of the following result. We give an outline of the proof.

Proposition 3.1 [8]. Let R be a right and left Noetherian ring and I an ideal of R. Then I is Artinian as a left module if and only if I is Artinian as a right module.

Proof. We suppose that ${}_RI$ has finite length. Using Noetherian induction we may also suppose that I is a minimal non-zero two-sided ideal. In this case, $P = r(I)$ is a prime ideal and, by Goldie's Theorem, $S = R/P$ has a simple Artinian quotient ring Q. We show that $S = Q$ and then the result follows easily. Note that I is a faithful S-module. Put $Z(I) = \{i \in I \mid \mathrm{ann}_S(I) \text{ is an essential right ideal of } S\}$. Then $Z(I) = 0$ or $Z(I) = I$. However, the latter case implies the existence of a regular element c of S such that $Ic = 0$, contradicting the faithfulness of I_S. Hence, for each regular element c of S, the endomorphism of ${}_RI$ given by $i \to ic$ is a monomorphism, and therefore an isomorphism, since ${}_RI$ has finite length. Thus $I = Ic$ and I is a torsion free, divisible S-module. Consequently I is also a Q-module. The fact that I is a Noetherian S-module implies that Q is a Noetherian S-module, and from this it follows easily that $S = Q$. ∎

Thus, in a right and left Noetherian ring, $\tau_0(R) = \tau_0({}_RR)$ and we write $\tau_0(R)$ for this ideal.

In order to obtain the desired result, we follow the general line of the previous section taking care to account for the lack of boundedness and symmetry. Throughout this section we assume that R is a right and left Noetherian ring.

Lemma 3.2. If R has a minimal prime P such that $|{}^R/P| = 0$, then $\tau_\circ(R) \neq 0$.

Proof. We first consider the case of a semiprime ring. In this case $\ell(P) \neq 0$ and obviously $\ell(P)$ is a right ${}^R/P$-module. Hence $\ell(P) \subseteq \tau_\circ(R)$.

In the non-semi-prime case we follow the proof of Theorem 2.4, using Proposition 3.1 instead of Theorem 2.1. ■

Corollary 3.3. If $\tau_\circ(R)$ is zero and N is the nilpotent radical of R then $\tau_\circ({}^R/P)$ and $\tau_\circ(R/\ell(N))$ are both zero.

Proof. The fact that $\tau_\circ(R/\ell(N))$ is zero follows from the left-hand version of Lemma 2.3.

Suppose that $\tau_\circ({}^R/N) \neq 0$. Since N is the intersection of the minimal prime ideals of R there is a minimal prime ideal P such that $\tau_\circ({}^R/N) \not\subseteq {}^P/N$. But then the prime ring ${}^R/P$ has a non-zero Artinian ideal and so ${}^R/P$ is itself an Artinian ring; that is, $|{}^R/P| = 0$. Using Lemma 3.2, we find a contradiction. ■

If I is an ideal of R, we put $'\mathcal{C}(I) = \{c \in R | rc \in I \Rightarrow r \in I\}$ and $\mathcal{C}(I) = \{c \in R | c + I$ is a regular element of $R/I\}$. If S is a non-empty set of elements of R we say that R satisfies the <u>left Ore condition</u> with respect to S if for each $r \in R$, $s \in S$ there exists $r' \in R$, $s' \in S$ such that $s'r = r's$.

Lemma 3.4. Let R be a right and left Noetherian ring such that $|R_R| = 1$ and $\tau_\circ(R) = 0$. If N is the nilpotent radical of R than $\mathcal{C}(N) = {}'\mathcal{C}(0)$ and R satisfies the left Ore condition with respect to this set.

Proof. It is enough to show that $\mathcal{C}(N) \subseteq {}'\mathcal{C}(0)$. For then the conclusion follows from a general result on Noetherian rings [3, 2.5].

The proof follows the same line as that of Theorem 2.6. ∎

In the proof of Theorem 2.6, once we have shown that $\mathcal{C}(N) \subseteq {}'\mathcal{C}(0)$ we appeal to symmetry in order to obtain $\mathcal{C}(N) = \mathcal{C}(0)$. However, we are not demanding that $|_R R| = 1$ and so we must show that Lemma 3.4 already contains enough information. In order to do this we need the following result due to P. F. Smith [9, Lemma 1.8].

Lemma 3.5. Suppose that R is a right and left Noetherian ring with nilpotent radical N. Suppose also that $\mathcal{C}(N) = {}'\mathcal{C}(0)$. Let $J = \{a \in R | ca = 0 \text{ for some } c \in \mathcal{C}(N)\}$. Then

(i) J is an ideal of R, and

(ii) there exists $c \in \mathcal{C}(N)$ such that $cJ = 0$.

Theorem 3.6. If R is a right and left Noetherian ring such that every non-zero right ideal has Krull dimension one then R has a two-sided Artinian quotient ring.

Proof. By Lemma 3.4, we know that $\mathcal{C}(N) = {}'\mathcal{C}(0)$, thus in order to use Small's Theorem we need only show that ${}'\mathcal{C}(0) = \mathcal{C}(0)$. If this is not true then the ideal J constructed in Lemma 3.5 is non-zero, and for some $c \in {}'\mathcal{C}(0)$, $cJ = 0$. Let P be an ideal of R maximal with respect to containing c and having a non-zero right annihilator. Then P is a prime ideal. Now $cR + N \subseteq P$ and $|R/(cR + N)| = 0$. Hence $|R/P| = 0$ and $|r(P)| = 0$, since $r(P)$ is a left R/P-module. Thus $\tau_0(R)$ is non-zero, a contradiction. ∎

Corollary 3.7 [c.f. 4, corollary 5]. If R is a right and left Noetherian ring with $|R_R| = 1$ and $\tau_0(R) = 0$ then a right ideal I of R contains a regular element if and only if $|R/I| = 0$.

Proof. ⇒ Use Lemma 1.4.

⇐ If $|R/I| = 0$ then $|R/(I + N)| = 0$. However, if I + N is not essential over N, this means that $\tau_0(R/N)$ is non-zero, contradicting Corollary 3.3. Hence I + N is essential over N and, by Goldie's Theorem, I contains an element of $\mathcal{C}(N)$, which is regular, by Theorem 3.6. ■

§4. Jacobson's Conjecture

Recently Cauchon [1], has shown that, in a left Noetherian, right fully bounded Noetherian ring, the intersection of the powers of the Jacobson radical is zero. In [7], Jategaonkar proves this result for fully bounded Noetherian rings. Here, using a combination of the methods of Cauchon and Jategaonkar, together with Proposition 3.1, we give a short transparent proof of this result. S. M. Ginn and P. B. Moss [2], have made the following observation.

Theorem 4.1. Let R be a right and left Noetherian ring whose right socle is right essential. Then R is Artinian.

Proof. Let S be the right socle. By Proposition 3.1, ${}_RS$ is left Artinian, and it follows that $R/\ell(S)$ is an Artinian ring. However, since $\ell(S).S = 0$, we see that $\ell(S)$ is in the singular ideal of R and so is nilpotent. Hence R is Artinian. ■

Theorem 4.2. If R is left Noetherian, right fully bounded Noetherian and J is the Jacobson radical of R, then $\bigcap_{n=1}^{\infty} J^n = 0$.

Proof. Suppose that $X = \bigcap_{n=1}^{\infty} J^n \neq 0$. By Noetherian induction we may assume that X is contained in every non-zero ideal of R. Let $I \subsetneq X$ be a right ideal such that X/I is simple. If $P = \mathrm{ann}_R(X/I)$ then R/P is simple Artinian, by Proposition 1.6.

Choose $I \subseteq L$ a right ideal maximal among right ideals such that $X \nsubseteq L$. Note that L does not contain a non-zero ideal. Put K = X + L. Then R/L is an essential extension the simple module K/L and $\mathrm{ann}_R(K/L) = P$.

In an arbitrary right ideal of R choose $0 \neq U$ a right ideal such that $Q = r(U)$ is a prime ideal. Then $RU + L \not\subseteq L$, and so $K \subseteq RU + L$. If $K = RU + L$ then $RUP \subseteq L$ and so $UP = 0$ and $Q = P$. Otherwise, $K \subsetneq RU + L$, and, since $(RU + L)Q \subseteq L$ we see that $(RU + L)/K$ is a singular R/Q-module. Hence, by Lemma 1.5, there exists $Q \subsetneq A$, an ideal of R with $(RU + L)A \subseteq K$. But then $(RU)AP \subseteq L$, which implies that $UAP = 0$ and, again, $Q = P$. In eigher case U is an R/P-module and therefore is Artinian. Thus R has essential socle and, by Theorem 4.1, is Artinian. As a consequence, J is nilpotent, a contradiction. ■

References

1. G. Cauchon, "Sur l'intersection des puissances du radical d'un T-anneau noethérian," C. R. Acad. Sc. Paris, 279 Série A, 91-93.

2. S. M. Ginn & P. B. Moss, "Finitely embedded modules over Noetherian rings," Bull. Amer. Math. Soc. 81(1975), 709-710.

3. A. W. Goldie, "The structure of Noetherian rings, Lectures on rings and modules," Springer Lecture Notes 246, 1972.

4. R. Gordon, "Artinian quotient rings of FBN rings," J. of Algebra 35(1975), 304-307.

5. __________, "Primary decomposition in right Noetherian rings," Communications in Algebra, 2(1974), 491-524.

6. __________ and J. C. Robson, "Krull dimension," Amer. Math. Soc. Memoirs, 133(1973).

7. A. V. Jategaonkar, "Jacobson's conjecture and modules over fully bounded Noetherian rings," J. of Algebra 30, 103-121.

8. T. H. Lenagan, "Artinian ideals in Noetherian rings," Proc. Amer. Math. Soc. 51(1975), 499-500.

9. P. F. Smith, "On two-sided Artinian quotient rings," Glasgow Mathematical Journal 13, 159-163.

CYCLIC AND FAITHFUL OBJECTS IN QUOTIENT CATEGORIES WITH APPLICATIONS TO NOETHERIAN SIMPLE OR ASANO RINGS

J. C. Robson
University of Leeds
Leeds, Great Britain

It is proved in [1, Lemma 3.1] that over a simple nonartinian ring, every module of finite composition length is cyclic. The key to this result is that, over a simple ring, every nonzero module is faithful.

In [13], this result is expanded to prove that over a simple Noetherian ring of Krull dimension > n, every finitely generated module of Krull dimension n has a generating set of n + 1 elements. It follows, for example, that in the Weyl algebra $A_n(k)$, where k is a field of characteristic zero, every right ideal has a generating set of n + 1 elements.

The main purpose of this paper is to describe the heuristic argument which suggested this expanded result, and to convert it into a formal method of proof. We hope that this method will be applicable more widely. Here we use it to reprove the above results and to generalize some of them to Noetherian rings whose ideals are progenerators, i.e., Noetherian Asano rings [9]. This generalization has the merit of yielding the like result for ideals in a Dedekind domain. More usefully, it also includes the polynomial ring in one variable over $A_n(k)$.

The heuristic argument is based upon the fact that a noetherian module, in an appropriate quotient category, becomes of finite length. Thus a result which can be proved for modules of finite length could also hold, in some form, for noetherian modules.

In the case in question, let R be a simple noetherian ring of Krull dimension > n and A a noetherian R-module of Krull dimension n. Let $\mathcal{A}$ be the category of right R-modules, and let $\mathcal{C}$ be the localizing subcategory of all modules of Krull dimension < n. Let $T:\mathcal{A} \to \mathcal{A}/\mathcal{C}$ be the canonical functor to the quotient category. By definition of Krull dimension [2], TA is an object of finite length. Since R is a simple ring, one would expect nonzero objects of $\mathcal{A}/\mathcal{C}$ to be "faithful". The proof of [1, Lemma 3.1] would, hopefully, then show that TA is "cyclic". In $\mathcal{A}$, this should mean that A has a cyclic submodule B such that $A/B \in \mathcal{C}$; and an inductive argument on the dimension n would show that A needs no more than n + 1 generators.

It is, we hope, clear from the quotation marks where this argument is lacking. We know of no established theory of faithful objects and cyclic objects in the quotient category other than, perhaps, the obvious one of regarding these objects as modules over the localization ring of R with respect to $\mathcal{C}$.

In the first two sections of this paper, we define these two concepts in such a way as to suit the results at which we aim. We also prove some of the elementary properties of cyclic and faithful objects.

In section 3, the question of how categorical these definitions are is discussed, together with a brief study of a special property of some localizing subcategories--namely, being closed under tensoring with ideals of the ring, or ideal invariance as we term it.

Finally in the last two sections we follow through the no-longer-heuristic argument outlined above and obtain results concerning generation of right ideals in simple noetherian rings and noetherian Asano rings.

I would like to thank the organizers of the conference at Kent State University for their invitation, organization and hospitality. I also want to thank Robert Gordon for spending many hours discussing with me the material in this paper; and Muriel Gordon for her fine patience during those many hours.

§1. Cyclicity.

Throughout this paper, we let R be a ring, $\mathbb{A}$ = mod-R, the category of right R-modules, $\mathcal{C}$ a localizing subcategory, and T: $\mathbb{A} \to \mathbb{A}/\mathcal{C}$ the canonical functor to the quotient category--see [2 or 14] for details. If $A \in \mathbb{A}$ we let $A_{\mathcal{C}}$ denote the largest submodule of A belonging to $\mathcal{C}$. We will be using the following result which comes from [4, Lemma 1.1].

Lemma 1.1. Let

$$0 \to TA \to TB \to TC \to 0$$

be a short exact sequence in $\mathbb{A}/\mathcal{C}$. Then there is a short exact sequence

$$0 \to A' \to B \to C' \to 0$$

in $\mathbb{A}$ together with isomorphisms $TA \to TA'$, $TC \to TC'$ making this diagram commutative,

$$\begin{array}{ccccccccc} 0 & \to & TA & \to & TB & \to & TC & \to & 0 \\ & & \downarrow & & \| & & \downarrow & & \\ 0 & \to & TA' & \to & TB & \to & TC' & \to & 0 \end{array}$$

As a preliminary to the definition of a cyclic object of A/C we have

Definition 1.2. A module $A \in \mathbb{A}$ is $\mathcal{C}$-cyclic if A has a cyclic submodule B such that $A/B \in \mathcal{C}$; that is, if there exists $f \in Hom(R,A)$ with Tf an epimorphism. We then call f a $\mathcal{C}$-generator of A.

Lemma 1.3. Let $A \in \mathbb{A}$. Then A is $\mathcal{C}$-cyclic if and only if $A/A_{\mathcal{C}}$ is $\mathcal{C}$-cyclic.

Proof. =>. Let f be a $\mathcal{C}$-generator of A, π the projection of A onto $A/A_{\mathcal{C}}$. Clearly πf is a $\mathcal{C}$-generator of $A/A_{\mathcal{C}}$.

<=. Let g be a $\mathcal{C}$-generator of $A/A_{\mathcal{C}}$. Lift g to $f \in Hom(R,A)$, so that $\pi f = g$. Since $T\pi$ is an isomorphism and Tg an epimorphism, Tf must be an epimorphism. ∎

We intend to work with cyclic objects of $\mathbb{A}/\mathfrak{C}$. Yet it is easy to see that there are modules A, B with TA $\cong$ TB and A being $\mathfrak{C}$-cyclic, B not. For example, let R = k[x, y] be the polynomial ring in two commuting indeterminates over a field k, and let P be the maximal ideal (x, y). Let $\mathfrak{C}$ be the localizing subcategory generated by the simple module R/P. Of course R is $\mathfrak{C}$-cyclic and TR $\cong$ TP. Yet P cannot be $\mathfrak{C}$-cyclic or else P would be minimal over a principal ideal.

We avoid this awkwardness by considering all the inverse images of TA; i.e., all modules B such that TA $\cong$ TB.

Definition 1.4. An object TA of $\mathbb{A}/\mathfrak{C}$ is cyclic if every inverse image of TA is $\mathfrak{C}$-cyclic.

The preceding example shows that even TR may not be cyclic. The next few results establish some basic facts concerning the behavior of cyclic objects under morphisms, change of rings, etc.. The first provides a source of cyclic objects.

Proposition 1.5. A simple object TA of $\mathbb{A}/\mathfrak{C}$ is cyclic.

Proof. Let B be an inverse image. Choose $0 \neq b \in B - B_{\mathfrak{C}}$. Then the map induced by $1 \to b$ is a $\mathfrak{C}$-generator of B. ∎

The next result provides a test for cyclicity.

Lemma 1.6. For $A \in \mathbb{A}$, the following are equivalent.

(i) TA is cyclic.

(ii) All $B \subseteq A$ such that $A/B \in \mathfrak{C}$ are $\mathfrak{C}$-cyclic.

Proof. We need only prove that (ii) => (i). So, suppose $D \in \mathbb{A}$ and TD $\cong$ TA. Then $\exists f \in \mathrm{Hom}(D', A/A')$ with D/D', A', ker, cokes $f \in \mathfrak{C}$. It will suffice to show that D' is $\mathfrak{C}$-cyclic. By hypothesis, $D'/\ker$ is $\mathfrak{C}$-cyclic. But then Lemma 1.3 implies that D' is $\mathfrak{C}$-cyclic. ∎

Lemma 1.7. Morphic images of cyclic objects are cyclic.

Proof. Let $Tg:TA \to TB$ be an epimorphism, with TA cyclic. We need to show that B is $\mathcal{C}$-cyclic. By Lemma 1.3, we may suppose that $B_{\mathcal{C}} = 0$. Thus we may assume that $g \in \mathrm{Hom}(A', B)$ where $A/A' \in \mathcal{C}$. Now A' is $\mathcal{C}$-cyclic; say it is $\mathcal{C}$-generated by f. Then $gf \in \mathrm{Hom}(R,B)$ is a $\mathcal{C}$-generator of B. ∎

Next we consider the situation when the ring decomposes as a finite product of rings, $R = \Pi R_i$, $i = 1, \ldots, n$. If we let $\mathcal{A}_i = \text{mod-}R_i$, then $\mathcal{A} = \Pi\mathcal{A}_i$ and $\mathcal{C} = \Pi\mathcal{C}_i$ with $\mathcal{C}_i$ being a localizing subcategory of $\mathcal{A}_i$. We let $T_i:\mathcal{A}_i \to \mathcal{A}_i/\mathcal{C}_i$ be the canonical functors.

Lemma 1.8. Let $TA = \Pi T_iA_i$. Then TA is cyclic if and only if every T_iA_i is cyclic.

Proof. =>. Given the A_i, let $A = \Pi A_i$. Let $f \in \mathrm{Hom}(R,A)$ be a $\mathcal{C}$-generator. Then the composite map

$$R_i \to R \xrightarrow{f} A \to A_i$$

is a $\mathcal{C}_i$-generator of A_i.

<=. Given A, decompose it as $A = \Pi A_i$. Let $f_i \in \mathrm{Hom}(R_i,A_i)$ be a $\mathcal{C}_i$-generator. Then $f \in \mathrm{Hom}(R,A)$ given by

$$f(r) = f(r_1, r_2, \ldots, r_n) = (f_1(r_1), f_2(r_2), \ldots, f_n(r_n))$$

is a $\mathcal{C}$-generator of A. ∎

Let $\bar{R}$ be a homomorphic image of the ring R. We may regard $\bar{\mathcal{A}} = \text{mod-}\bar{R}$ as a subcategory of $\mathcal{A}$. Let $\bar{\mathcal{C}} = \mathcal{C} \cap \bar{\mathcal{A}}$ and let $\bar{T}:\bar{\mathcal{A}} \to \bar{\mathcal{A}}/\bar{C}$ be the canonical functor.

Lemma 1.9. Let $A \in \bar{\mathcal{A}}$. Then TA is cyclic if and only if $\bar{T}A$ is cyclic.

Proof. =>. Any inverse image B of $\bar{T}A$ in $\bar{\mathcal{A}}$ is also an inverse image of TA. Hence B is $\mathcal{C}$-cyclic and so $\bar{\mathcal{C}}$-cyclic.

<=. We need to show that if B is an inverse image of TA in $\mathbb{A}$ then B is $\mathcal{C}$-cyclic. By Lemma 1.6 we may suppose that $B \subseteq A$. But then $B \in \bar{\mathbb{A}}$ and so is an inverse image of $\overline{T}A$. Therefore B is $\overline{\mathcal{C}}$-cyclic and hence $\mathcal{C}$-cyclic. ■

The final result in this section is a version of Nakayama's lemma. In order to prove it we use an additional hypothesis on the localizing subcategory $\mathcal{C}$.

<u>Definition 1.10</u>. We call $\mathcal{C}$ <u>ideal</u> <u>invariant</u> if for every $C \in \mathcal{C}$ and every (two-sided) ideal X of R we have $C \otimes_R X \in \mathcal{C}$

It seems wise to see immediately that not all localizing subcategories have this property.

<u>Example 1.11</u>. Let S be a ring having a maximal right ideal X which is not a two-sided ideal. Let $R = \mathrm{II}(x) = \{s \in S \mid sX \subseteq X\}$. Then X is an ideal of the idealizer ring R. Moreover S/R and R/X are nonisomorphic simple right R-modules, and

$$S/R \otimes_R X \cong S/X,$$

(see [11, §1 and 12, Lemma 2.2]). Thus if we let $\mathcal{C}$ be the localizing subcategory of $\mathbb{A}$ = mod-R generated by S/R it follows that $R/X \notin \mathcal{C}$ and so $S/R \otimes_R X \notin \mathcal{C}$. So $\mathcal{C}$ is not ideal invariant.

The question of when localizing subcategories are ideal invariant we prefer to discuss later, in section 3.

<u>Lemma 1.12</u>. Let $\mathcal{C}$ be ideal invariant, $A \in \mathbb{A}$, X an ideal of R, and n a natural number. Then T(A/AX) is cyclic if and only if $T(A/AX^n)$ is cyclic.

<u>Proof</u>. <=. This is immediate from Lemma 1.7.

=>. We may suppose, without loss, that $AX^n = 0$. Also, using Lemma 1.6, it will suffice to show that if $B \subseteq A$ and $A/B \in \mathcal{C}$ then B is $\mathcal{C}$-cyclic. We make the induction hypothesis that $T(A/AX^{n-1})$ is cyclic.

Note that $(B + AX^{n-1})/AX^{n-1}$ is an inverse image of $T(A/AX^{n-1})$. Let f be a $\mathcal{C}$-generator of $(B + AX^{n-1})/AX^{n-1}$ and let $g \in \text{Hom}(R, B)$ be a lifting of f. We write $g(1) = b$. Now $A/(bR + AX^{n-1}) \in \mathcal{C}$. Since $\mathcal{C}$ is ideal invariant,

$$(A/(bR + AX^{n-1})) \otimes X^{n-1} \in \mathcal{C}.$$

Therefore its image under the multiplication map, namely

$$AX^{n-1}/(bX^{n-1} + AX^{2n-2}) = AX^{n-1}/bX^{n-1},$$

belongs to $\mathcal{C}$. Hence $AX^{n-1}/(bR \cap AX^{n-1}) \in \mathcal{C}$ and so $(bR + AX^{n-1})/bR \in \mathcal{C}$. This, together with the fact that $A/bR + AX^{n-1} \in \mathcal{C}$, shows that $A/bR \in \mathcal{C}$; and so B is $\mathcal{C}$-cyclic. ∎

§2. Fidelity.

Our aim in this section is to define annTA for $A \in \mathcal{A}$, and to establish basic properties of this annihilator. Of course we again have the problem that quite different modules in $\mathcal{A}$ become isomorphic in $\mathcal{A}/\mathcal{C}$. For example, using the notation of Example 1.11, if $A = R/X$ and $B = S/X$ then $TA \cong TB$ and yet $\text{ann}A = X$, $\text{ann}B = 0$. In fact $X \notin \mathcal{C}$ and so even $T(\text{ann}A) \neq T(\text{ann}B)$.

Definition 2.1. For $TA \in \mathcal{A}/\mathcal{C}$ we define

$$\text{ann } TA = \textstyle\sum_i \{\text{ann } B_i \mid TB_i \cong TA\}.$$

This clearly has the advantage of being preserved under isomorphism in $\mathcal{A}/\mathcal{C}$.

Lemma 2.2. Let $A \in \mathcal{A}$.

(i) $\text{ann } TA = \text{ann } T(A/A_{\mathcal{C}})$.

(ii) If $A_{\mathcal{C}} = 0$, then $\text{ann } TA = \sum_i \{\text{ann } B_i \mid B_i \subseteq A,\ A/B_i \in \mathcal{C}\}$.

(iii) If $A_{\mathcal{C}} = 0$ and R has the a.c.c. on ideals, then $\exists B \subseteq A$ with $A/B \in \mathcal{C}$ and $\text{ann } TA = \text{ann } B$.

Proof. (i) $TA \cong T(A/A_{\mathcal{C}})$.

(ii) We need to show that if $TD \cong TA$ then $\text{ann } D \subseteq \sum \text{ann } B_i$. Since $TD \cong TA$, $\exists f \in \text{Hom}(D', A/A')$ with D/D', A', ker, coker $f \in \mathcal{C}$. But $A_{\mathcal{C}} = 0$, so $A' = 0$. Let $\text{im } f = B \subseteq A$. Then $A/B \in \mathcal{C}$ and

$$\text{ann } D \subseteq \text{ann } D' \subseteq \text{ann } B \subseteq \textstyle\sum \text{ann } B_i.$$

(iii) If B_i, B_j are as in (ii), then $B_i \cap B_j \subseteq A$ and $A/B_i \cap B_j \in \mathcal{C}$. Thus $B_i \cap B_j = B_k$ for some k. Hence if B is chosen amongst the B_i so as to maximize ann B, then

$$\text{ann } B = \text{ann } (B \cap B_j) \supseteq \text{ann } B_j$$

for all j. Hence $\text{ann } TA = \text{ann } B$. ∎

<u>Lemma 2.3</u>. $R_{\mathcal{C}} \subseteq \text{ann } TA$.

<u>Proof</u>. $A.R_{\mathcal{C}} \subseteq A_{\mathcal{C}}$. Thus

$$R_{\mathcal{C}} \subseteq \text{ann } T(A/A_{\mathcal{C}}) = \text{ann } TA. \blacksquare$$

In view of this result, we define TA to be <u>faithful</u> if ann TA $= R_{\mathcal{C}}$; equivalently, if T ann TA = 0. Otherwise we say TA is <u>unfaithful</u>.

<u>Lemma 2.4</u>. If

$$0 \to TA \to TB \to TC \to 0$$

in a short exact sequence in $\mathcal{A}/\mathcal{C}$ then ann TB $\subseteq$ ann TA and ann TB $\subseteq$ ann TC.

<u>Proof</u>. It suffices to show that if ann B = X then X $\subseteq$ ann TA and X $\subseteq$ ann TC. By Lemma 1.1, there is a corresponding short exact sequence

$$0 \to A' \to B \to C' \to 0$$

with $TA \cong TA'$ and $TC \cong TC'$. Clearly

$$X \subseteq \text{ann } A' \subseteq \text{ann } TA' = \text{ann } TA$$

and likewise

$$X \subseteq \text{ann } TC. \blacksquare$$

We will need a converse result later--but one which requires the extra hypothesis that $\mathcal{C}$ is ideal invariant. As a preliminary we prove

<u>Lemma 2.5</u>. If $\mathcal{C}$ is ideal invariant then

$$A \text{ ann } TA \subseteq A_{\mathcal{C}}.$$

<u>Proof</u>. We may suppose $A_{\mathcal{C}} = 0$. Then, by Lemma 2.2(ii), ann TA = $\sum\{\text{ann } B_i \mid B_i \subseteq A, A/B_i \in \mathcal{C}\}$. Let ann $B_i = X_i$. Then

$$(A/B_i) \otimes X_i \cong A \otimes X_i/B_i \otimes X_i$$

which has AX_i as a homomorphic image via multiplication. Hence $AX_i \subseteq A_{\mathcal{C}}$ as required. ■

Lemma 2.6. If

$$0 \to TA \to TB \to TC \to 0$$

is a short exact sequence and $\mathcal{C}$ is ideal invariant then

$$(\text{ann } TC)(\text{ann } TA) \subseteq \text{ann } TB.$$

Proof. By Lemma 1.1, there is a corresponding short exact sequence

$$0 \to A' \to B \to C' \to 0.$$

We will identify A' with its image in B. Now, by Lemma 2.5, $C' \text{ ann } TC \subseteq C'_{\mathcal{C}}$. Hence

$$T(B \text{ ann } TC) \cong TA' \cong TA.$$

Therefore, using Lemma 2.5 again,

$$B(\text{ann } TC)(\text{ann } TA) \subseteq B_{\mathcal{C}}.$$

Hence, by Lemma 2.2(i), $(\text{ann } TC)(\text{ann } TA) \subseteq \text{ann } TB$. ■

Example 2.7. To see that the extra condition upon $\mathcal{C}$ was necessary, we consider again the situation described in Example 1.11. Let $A = C = S/X$ and let B be a nonsplit extension

$$0 \to A \to B \to C \to 0.$$

That such extensions can exist, even when S is a simple hereditary noetherian domain, is shown in [6, Corollary 5.20]. View these modules as R-modules. Note, by [11; Corollary 1.5] that B then has a unique composition series of length 4; say

$$B = B_4 \supset B_3 \supset B_2 \supset B_1 \supset B_0 = 0,$$

where $B_4/B_3 \cong B_2/B_1 \cong S/R$ and $B_3/B_2 \cong B_1/B_0 \cong R/X$. It is not difficult, using Lemma 2.2(ii), to check that ann TA = ann TC = X and ann TB = 0. However, by choice of X, SX = X and so

$$\text{ann TC ann TA} = X^2 = X \not\subseteq \text{ann TB}.$$

If we also note that $AX = A \not\subseteq A_{\mathcal{C}}$, we see that Lemma 2.5 also fails for general $\mathcal{C}$.

The fact, for $A \in \mathcal{A}$, that

$$\text{ann } A = \bigcap\{\ker|f \in \text{Hom}(R,A)\},$$

whilst elementary, is very useful in practice. A similar comment applies to the corresponding fact established in the next result, Note that the containment there cannot be replaced by an inequality, as is shown by the comments at the beginning of this section.

<u>Lemma 2.8</u>. If $A \in \mathcal{A}$ and $A_{\mathcal{C}} = 0$ then

$$\bigcap\{\ker Tf|f \in \text{Hom}(R,A)\} = T \text{ ann } A \subseteq T \text{ ann } TA.$$

<u>Proof</u>. Of course the final containment is obvious. Note also that, if $f \in \text{Hom}(R,A)$, then ann $A \subseteq \ker$ and so

$$T \text{ ann } A \subseteq T \ker = \ker Tf.$$

Hence $T \text{ ann } A \subseteq \bigcap\{\ker Tf|f \in \text{Hom}(R,A)\}$.

On the other hand $\bigcap \ker Tf \subseteq TR$; say $\bigcap \ker Tf = TB$, $B \subseteq R$. Then $TB \subseteq T\ker$ for all f. Therefore, for each f there is a right ideal $B_f \subseteq B$ such that $B_f \subseteq \ker$, and $B/B_f \in \mathcal{C}$. However f(B) is a homomorphic image of B/ker and so $f(B) \in \mathcal{C}$. Since $A_{\mathcal{C}} = 0$, $B \subseteq \ker$. Thus $B \subseteq \bigcap \ker$ and $TB \subseteq T \bigcap \ker$. ∎

As an easy consequence we have

<u>Corollary 2.9</u>. If $A \in \mathcal{A}$ and $A_{\mathcal{C}} = 0$ and TA is faithful, then $\bigcap\{(\ker Tf|f \in \text{Hom}(R,A)\} = 0$.

§3. Categorical Considerations.

In this section we concentrate on the following two questions. Which localizing subcategories are ideal invariant? How categorical are the notions of cyclic and faithful introduced above?

First we note that, to check if $\mathcal{C}$ is ideal invariant, it suffices to show that, for each cyclic module $A \in \mathcal{C}$ and each ideal X of R, we have $A \otimes X \in \mathcal{C}$. Suppose $A \cong R/I$. Then

$$A \otimes X \cong X/IX.$$

Hence we need only check that

$$R/I \in \mathcal{C} \Rightarrow X/IX \in \mathcal{C}.$$

<u>Lemma 3.1</u>. Every localizing subcategory is ideal invariant if

(i) R is simple or

(ii) R is commutative.

<u>Proof</u>. (i) is obvious.

(ii) Of course X/IX is an R/I-module. Since $R/I \in \mathcal{C}$ so too is X/IX. ■

We note that Example 1.11, when applied to the case where $S = M_n(k)$ $n \geq 2$, k a field, shows that for fully bounded noetherian rings (FBN-rings) not all localizing subcategories are ideal invariant.

However the localizing subcategories which concern us here are those involved in Gabriel's notion of Krull dimension (see [2, 3, 4, 7]). Let R be a noetherian ring and let $\mathcal{C}_\alpha$ be the smallest localizing subcategory of $\mathcal{A}$ containing all modules of Krull dimension $< \alpha$.

We recall that R is an <u>Asano ring</u> if its nonzero ideals are both left and right progenerators.

<u>Proposition 3.2</u>. If either

(i) R is a noetherian Asano ring, or

(ii) R is an FBN-ring

then the localizing subcategories $\mathcal{C}_\alpha$ are ideal invariant.

<u>Proof</u>. (i) Membership of $\mathcal{C}_\alpha$ is determined by the lattice of submodules of a module. Thus $\mathcal{C}_\alpha$ is invariant under a category equivalence of $\mathbb{A}$ with itself, and hence under tensoring with an ideal.

(ii) Let $R/I \in \mathcal{C}_\alpha$, and X be an ideal of R. By [5, Lemma 2.1], if $Y = \text{ann } R/I$ then $R/Y \in \mathcal{C}_\alpha$. Therefore, by [5, Lemma 2.2] the left Krull dimension of R/Y is less than α. Hence the same holds for X/YX. Applying [5, Lemma 2.2] again, we see that $X/YX \in \mathcal{C}_\alpha$. However X/IX is a homomorphic image of X/YX and so $X/IX \in \mathcal{C}_\alpha$. ∎

We now take up the second question--how categorical are the definitions of sections 1, 2? The obvious categorical setting for this question is a Grothendieck category $\mathcal{G}$ since, by the Gabriel-Popescu theorem (see [7]), such categories are precisely those of the form mod-R/$\mathcal{C}$. However $\mathcal{G}$ may be presented as a quotient category of many different module categories. One could, for example, vary the presentation by changing the choice of generator in $\mathcal{G}$. But, even without doing this, there are variations in presentation.

For example, if R is a noetherian prime ring, $\mathcal{C}$ the localizing subcategory of torsion modules and Q the classical quotient ring of R, then

$$\text{mod-}R/\mathcal{C} \simeq \text{mod-}Q.$$

with TR corresponding to Q under this equivalence.

As defined, our notions of being cyclic or faithful depend upon the particular presentation of $\mathcal{G}$ as $\mathbb{A}/\mathcal{C}$. To see that they are indeed dependent we consider again the situation of Example 2.7, taking S to be the simple domain. It can be seen that

$$\text{mod-}S \simeq \text{mod-}R/\mathcal{C} = \mathbb{A}/\mathcal{C}$$

with S corresponding to TR. Of course S/X is a faithful S-module yet, viewed as an object in $\mathcal{A}/\mathcal{C}$, $S/X = T(R/X)$ and ann $T(R/X) = X$. Since $TX \neq 0$ we see that S/X is faithful viewed as an object of mod-S but not when viewed as an object of $\mathcal{A}/\mathcal{C}$.

Likewise, if we consider the S-module $A = S/X \oplus S/X$ then, by [1, Lemma 3.1] A is cyclic. Yet, as an object of $\mathcal{A}/\mathcal{C}$, it cannot be cyclic. For $A = T(R/X \oplus R/X)$ which can be seen to be noncyclic since $R/X \oplus R/X$ is not cyclic.

These examples establish that our notions are concerned with a particular presentation of $\mathcal{G}$. However there is one related result which will be needed later. Suppose that R is a noetherian prime ring, C the localizing subcategory of torsion modules Q the classical quotient ring of R. So $\mathcal{A}/\mathcal{C} \cong$ mod Q. We identify both as $\mathcal{G}$.

Lemma 3.3. The following conditions on an object of $\mathcal{G}$ are equivalent.

(i) It is cyclic as a Q-module.

(ii) It has length $\leqslant$ length of Q.

(iii) It is cyclic as an object of $\mathcal{A}/\mathcal{C}$.

Proof. Note first that a Q-module, D say, is cyclic if and only if length $D \leq$ length Q. Moreover, if length $D = d$, then D will be a direct sum of d minimal right ideals of Q. Let $D = TA$ viewed as an object of $\mathcal{A}/\mathcal{C}$. Then we aim to show that A is $\mathcal{C}$-cyclic. By Lemma 1.3, we may assume that $A_{\mathcal{C}} = 0$. Then A will contain a submodule B with $A/B \in \mathcal{C}$ and B a direct sum of d uniform right ideals of R. The existence of regular elements in essential right ideals enables one to see that B is $\mathcal{C}$-cyclic--and so too, then, is A.

Conversely if A is $\mathcal{C}$-cyclic the $D = TA$ is an epimorphic image of $TR = Q$; and so D is cyclic. ■

§4. Simple Rings.

Having defined and established the elementary properties of cyclic objects and annihilators of objects in $\mathcal{A}/\mathcal{C}$ we are now in a position to extend the result of [1, Lemma 3.1] concerning modules of finite length. This we do in the next two results.

Proposition 4.1. Suppose there is a nonsplit short exact sequence

$$0 \to TA \xrightarrow{T\alpha} TB \xrightarrow{T\beta} TC \to 0$$

in $\mathcal{A}/\mathcal{C}$ where TC is cyclic and TA is simple. Then TB is cyclic.

Proof. It is enough to prove that B is $\mathcal{C}$-cyclic. Using Lemma 1.1, we choose a corresponding short exact sequence in $\mathcal{A}$,

$$0 \to A' \to B \to C' \to 0.$$

Let $f \in \mathrm{Hom}(R,C')$ be a $\mathcal{C}$-generator and let $g \in \mathrm{Hom}(R,B)$ be a lifting of f. Then $T\beta Tg$ is an epimorphism and so $\mathrm{im}T\alpha + \mathrm{im}Tg = TB$. Now it cannot be that $\mathrm{im}T\alpha \cap \mathrm{im}Tg = 0$ or else $T\beta$, restricted to imTg, is an isomorphism and the sequence splits. Yet $\mathrm{im}T\alpha \cong TA$ which is simple. Thus $\mathrm{im}T\alpha \cap \mathrm{im}Tg = \mathrm{im}T\alpha$ and so $\mathrm{im}Tg = TB$. By definition, B is $\mathcal{C}$-cyclic. ∎

Proposition 4.2. Let

$$0 \to TA \to TB \to TC \to 0$$

be a short exact sequence in $\mathcal{A}/\mathcal{C}$. Suppose that TC is cyclic and of finite length, that TA is simple, and that T(R/annTA) is not of finite length. Then TB is cyclic.

Proof. By Proposition 4.1, we can restrict our attention to the case when the sequence is split. It is easily seen, using Lemmas 1.1 and 1.3 that we need only prove that $A \oplus C$ is $\mathcal{C}$-cyclic. Choose $g \in \mathrm{Hom}(R,C)$ which $\mathcal{C}$-generates C. Since T(R/annTA) is not of finite length,

$$T(\text{ann}TA) \not\supseteq \ker Tg.$$

By Lemma 2.8, $\exists f \in \text{Hom}(R,A)$ with $\ker Tf \not\supseteq \ker Tg$. We let $h = (f,g) \in \text{Hom}(R, A \oplus C)$. Clearly $\text{im } Th = T(A \oplus C)$, thus h is a $\mathcal{C}$-generator of $A \oplus C$. ∎

We can now deduce Stafford's results [13].

<u>Theorem 4.3</u>. Let R be a simple Noetherian ring of finite Krull dimension, n say.

(i) If A is a Noetherian module and $\text{Kdim}A = m < n$ then A has a generating set of $m + 1$ elements.

(ii) Every right ideal of R has a generating set of $n + 1$ elements.

<u>Proof</u>. (i). Let $\mathcal{C}_m$ be the smallest localizing subcategory containing all modules of Krull dimension $< m$. Then TA has finite length (see [4]). Note that, since R is simple, every nonzero object in $\mathcal{A}/\mathcal{C}_m$ has zero annihilator. Thus a simple induction on length, using Proposition 4.2 and based upon Proposition 1.5, shows that TA is cyclic. Therefore, there is a cyclic submodule B of A with $T(A/B) = 0$. Thus $A/B \in \mathcal{C}_m$. By induction we may suppose that A/B has a set of m generators and so A has $m + 1$.

(ii). Let I be a right ideal of R. Choose any element $x \in I$ such that xR is an essential submodule of I. Then, by [3, Proposition 6.1], $\text{Kdim}\,(I/xR) \leq n - 1$. The result follows from (i). ∎

As a response to a question raised at the conference by Carl Faith we give

<u>Theorem 4.4</u>. The following conditions upon a ring R are equivalent:

(i) Every right module of finite length is cyclic.

(ii) Every left module of finite length is cyclic.

(iii) No factor ring is right artinian.

(iv) No factor ring is left artinian.

Proof. (iii) ⇒ (i). This follows almost immediately from Proposition 4.2.

(iii) ⇔ (iv). If R had a right artinian factor ring then it would have a simple artinian factor ring and vice versa.

(i) ⇒ (iii). Suppose R has a right artinian factor ring R/X. Then the module R/X ⊕ R/X if cyclic would be a homomorphic image of R/X. This is impossible since it has length greater than that of R/X. ∎

§5. Asano Rings.

In this section we extend some of the results of Section 4 to the case of a noetherian Asano ring. Such a ring is automatically prime and some characterizations and properties of noetherian Asano rings can be found in [9, 10].

Throughout this section R will be a noetherian Asano ring and $\mathcal{C}$ will be one of the $\mathcal{C}_\alpha$ which, by Proposition 3.2, is ideal invariant.

<u>Proposition 5.1</u>. Suppose that TR is not artinian and that

$$0 \to TA \to TB \to TC \to 0$$

is a short exact sequence with TA simple unfaithful and TC simple faithful. Then the sequence splits.

<u>Proof</u>. Suppose false. Then, by Proposition 4.1, TB is cyclic. We choose B such that B is cyclic and $B_{\mathcal{C}} = 0$. Then, by Lemma 1.1, we have a short exact sequence

$$0 \to A \to B \to C \to 0$$

for some choice of A and C.

Let $B \cong R/I$, $A \cong J/I$, $C \cong R/J$ where I, J are right ideals of R. By Lemma 2.5, ann TA = ann A = X say. We consider the modules involved in this diagram.

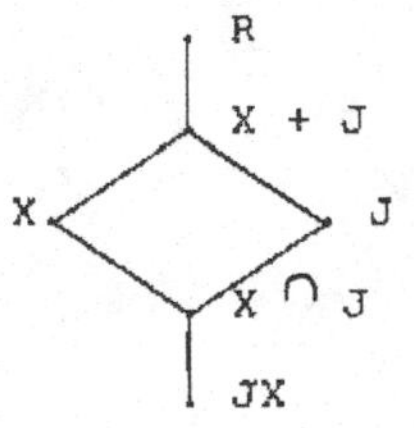

Since T(R/J) is faithful, $(R/J)X = X + J/J \notin \mathcal{C}$. Yet T(R/J) is simple. Thus

$$T(R/J) \cong T(X + J/J) \cong T(X/X \cap J)$$

But, if we let $\mathcal{L}$ denote the lattice of submodules of a module, then

$$\mathcal{L}(R/J) \cong \mathcal{L}((R/J) \otimes X)) \cong \mathcal{L}(X/JX)$$

since X is a left and right R-progenerator. Hence T(X/JX) is simple and so $T((X \cap J)/JX) = 0$.

Therefore

$$T(R/JX) \cong T(R/X) \oplus T(R/J).$$

Note that $J/I \cong A$ and $X = \text{ann } A$. Therefore $JX \subseteq I$. Hence

$$T(R/I) \cong T(R/X + I) \oplus T(R/J)$$

which shows that TB is split. []

Corollary 5.2. Suppose that TR is not artinian and that TB is an object of finite length. If TB is faithful then TB has a faithful simple subobject.

Proof. We proceed via induction on the length of TB. Let TA be a simple subobject. If it is faithful we are done. So suppose TA is unfaithful. Using Lemma 2.6, and remembering that R is prime, we see that T(B/A) is faithful. By our induction hypothesis, T(B/A) thus has a faithful simple subobject. However, by Proposition 5.1, the extension of TA by T(B/A) splits; and so TB has a faithful simple subobject. ■

This corollary, together with Proposition 4.2, shows that cyclicity of an object of finite length is going to depend upon its unfaithful factor objects.

Proposition 5.3. Let TR not be artinian. Let TA be an unfaithful object of finite length contained in a morphic image of TR. Then TA is cyclic.

Proof. We are concerned with the situation

$$0 \to TA \to T(R/I) \to TB \to 0$$

in which we choose I so that $(R/I)_{\mathcal{C}} = 0$. Then there is a corresponding short exact sequence which we may write as

$$0 \to A \to R/I \to B \to 0.$$

Say $A \cong J/I$ and thus $B \cong R/J$. Note that, if ann TA = X then $JX \subseteq I$, using Lemma 2.5. Thus, in order to conclude that TA is cyclic it will suffice, using Lemma 1.7, that T(J/JX) is cyclic.

To see this, recall that, since R is a Noetherian Asano ring, X has a unique factorization as a commutative product of maximal ideals; and $R/X = \bar{R}$ then decomposes as a finite direct product. By Lemma 1.9, and in its notation, it will suffice to prove that $\bar{T}(J/JX)$ is $\bar{\mathcal{C}}$-cyclic. By Lemma 1.8 we may suppose that $\bar{R}$ is indecomposable--which amounts to saying that X is a power of a maximal ideal. But then, using Lemma 1.12, we can actually assume X to be maximal.

Consider, then, $T(\bar{R})$. If this has not got finite length then nor has $\bar{T}(\bar{R})$. Applying Proposition 4.2 to the $\bar{R}$-module J/I, remembering that $\bar{R}$ is simple, we see that $\bar{T}(J/I)$ is cyclic. Hence, by Lemma 1.9, T(J/I) is cyclic.

On the other hand, suppose $T(\bar{R})$ does have finite length. Then it must be that K dim $(\bar{R}) = \alpha$ where $\mathcal{C} = \mathcal{C}_\alpha$. Thus $\bar{\mathcal{A}}/\bar{\mathcal{C}} \cong$ mod-Q where Q is the classical quotient ring of $\bar{R}$, using [3, Proposition 6.1]. Choose J' with $J \subseteq J' \subseteq R$ such that T(J'/J) is the largest artinian subobject of T(R/J). The facts that $T(X \cap J'/J'X)$ is artinian and $\mathcal{L}(T(X/J'X)) \cong \mathcal{L}((T(R/J'))$ show that $T(X \cap J') = T(J'X)$. Moreover length T(J'/J) = length T(J'X/JX). Hence

$$\begin{aligned} \text{length } T(J/JX) &= \text{length } T(J'/J'X) \\ &= \text{length } T(J' + X/X) \\ &\leq \text{length } T(R/X) \end{aligned}$$

and so, by Lemma 3.3, T(J/JX) is cyclic. ∎

Theorem 5.4. Let R be a noetherian Asano ring with K dim R = n, a natural number. Then every right ideal of R has a generating set of n + 1 elements.

Proof. This follows the same lines as that of Theorem 4.3, but relies upon Proposition 5.3 rather than Proposition 4.2. ■

It is known (see [8, 9]) that the polynomial ring in one commuting variable over a simple ring is Asano. Thus Theorem 5.4, together with [3, Theorem 9.2] yields

Corollary 5.5. Let S be a simple noetherian ring of finite Krull dimension, n say. Then every right ideal of the polynomial ring S[x] has a generating set of n + 2 elements.

This, in particular, applies to the case when S is a Weyl algebra. It raises a rather curious question. Namely, what conditions are imposed upon a ring S by the property that there is a bound on numbers of generators of right ideals in S[x]?

Finally it should be noted that Theorem 5.4 and Corollary 5.5 can be proved by methods similar to those used in [13]. This was shown jointly with J. T. Stafford (unpublished).

REFERENCES

1. D. Eisenbud and J. C. Robson, Modules over Dedekind prime rings, J. Algebra 16(1970), 67-85.

2. P. Gabriel, Des catégories abélliennes. Bull. Soc. Math. France 90(1962), 323-448.

3. R. Gordon and J. C. Robson, "Drull dimension", Memoir Amer. Math. Soc. 133(1973).

4. R. Gordon and J. C. Robson, The Gabriel dimension of a module, J. Algebra 29(1974), 459-473.

5. A. V. Jategaonkar, Jacobson's conjecture and moduels over fully bounded noetherian rings, J. Algebra, 30(1974), 103-121.

6. J. C. McConnell and J. C. Robson, Homomorphisms and extensions of modules over some differential operator rings, J. Algebra, 26(1973), 319-342.

7. N. Popescu, "Abelian categories with applications to rings and modules" Academic Press, 1973.

8. J. C. Robson, Pri-rings and ipri-rings, Oxford Quarterly J. or Math. 18(1967), 125-145.

9. ________, Noncommutative Dedekind rings, J. Algebra 9(1968), 249-265.

10. ________, A Note on Dedekind prime rings, Bull. London Math. Soc. 3(1971), 42-46.

11. ________, Idealizers and hereditary noetherian prime rings, J. Algebra 22(1972), 45-81.

12. ________, The coincidence of idealizer subrings, J. London Math. Soc. 10(1975).

13. J. T. Stafford, Completely faithful modules and ideals of simple noetherian rings, Bull. London Math. Soc. (to appear).

14. R. G. Swan, "Algebraic K-theory" Springer-Verlag Lecture Notes in Math. 76(1968).

PRIME SINGULAR-SPLITTING RINGS WITH FINITENESS CONDITIONS

Mark L. Teply
University of Florida
Gainesville, Florida

In this paper, all rings will be associative and have an identity element. Unless otherwise noted, all modules will be unital right modules.

A module M is said to split if its singular submodule Z(M) is a direct summand of M. If every R-module splits, R is called a splitting ring. Every splitting ring is a right nonsingular ring. In the last eight years, considerable progress has been made on the problem of characterizing all splitting rings. In a recent series of papers, Cateforis and Sandomierski [2], Goodearl [3, 4, 5, 6, 7] and Teply [13, 14] have reduced the problem of characterizing all splitting rings R to studying the case where R is a prime ring with zero right socle and homological dimension ≤ 2. In addition, Goodearl [8] has characterized the left and right Noetherian, hereditary, prime (HNP) splitting rings in the following ways: (1) R is a HNP splitting ring if and only if it is an iterated idealizer of a simple HNP splitting ring; (2) R is a HNP splitting ring if and only if R contains a minimal, nonzero, two-sided ideal and all faithful simple modules are injective. We note that, if I is the minimal, nonzero ideal of the HNP ring, then R/I is left and right Artinian. In an HNP splitting ring, every maximal ideal must be idempotent.

In this paper we propose to show that many of these properties of HNP splitting rings still hold when the "left and right hereditary" hypothesis is dropped and the "noetherian" hypothesis is weakened.

In section 1, we show that, if R is a prime splitting ring with zero right socle and if each two-sided ideal of R is finitely generated as a right ideal, then R has a minimal, nonzero, two-sided ideal I and

R/I is right Artinian with idempotent maximal ideals.

In section 2, we develop the theory of subidealizers. If T is a ring, and M is a right ideal of T, then the idealizer [10] of M in T is $\{x \in T | xM \subseteq M\}$. Note that the idealizer is a unital subring of T. A subidealizer R [8] of M is a unital subring of the idealizer which contains M; so M is a two-sided ideal of any subidealizer. We study the relationships of R and T with respect to semiprimeness, the Goldie conditions, and splitting. This leads to generalizations of results in [1, 6, 7].

A right ideal M of a ring T is called generative [7] if TM = T. A ring R is called an iterated subidealizer of a ring T if there exists a chain $R = R_0 \subset R_1 \subset R_2 \subset \ldots \subset R_n = T$ of rings such that each R_i $(i = 0,1,2, \ldots, n-1)$ is a subidealizer of an essential, generative right ideal of R_{i+1}.

In section 3, we apply the results of section 2 to obtain the following analogue of (1) for HNP rings: if R is a prime splitting ring with zero right socle and certain finiteness conditions, which are satisfied by all known examples of prime splitting rings, then R is an iterated subidealizer of a simple splitting ring with zero right socle and the same finiteness conditions. The results from section 2 also give us considerable information about the conditions under which the converse of this theorem holds.

We shall use the notation (0:X) to denote the right annihilator of a set X.

Most of the undefined ring-theoretic terminology can be found in [9]. Properties of singular modules are discussed in [3] and [12].

§1. Prime splitting rings for which two-sided ideals are finitely generated left ideals.

In this section we are interested in studying the two-sided ideals of a prime splitting ring with zero right socle. All the known examples of such prime rings have the property that the two-sided ideals are finitely generated as left ideals.

The proof of the following lemma is essentially contained in the proof of [3, Theorem 5.3].

Lemma 1.1. If R is a splitting ring and I is a two-sided ideal of R, then R/I is a right perfect ring whenever I is essential as a right ideal of R.

Our first proposition shows that a prime splitting ring cannot have very many maximal two-sided ideals which are finitely generated as left ideals. We note that, in a prime ring R, every nonzero two-sided ideal is essential as a left and a right ideal of R.

Proposition 1.2. Let R be a prime splitting ring with zero right socle. Let A be the set of two-sided maximal ideals of R which are finitely generated as left ideals of R. Then A is a finite set (or the empty set).

Proof. Suppose that $\{M_i\}_{i=1}^{\infty}$ is a set of distinct elements of A. A submodule S of a module M will be called divisible if $SM_i = S$ for each M_i (i=1, 2, ...). Then 0 is the only divisible submodule of $P = \Pi_{i=1}^{\infty} R/M_i$. Note that $\oplus_{i=1}^{\infty} R/M_i$ is singular; so P/Z(P) is divisible since each M_i is finitely generated as a left ideal. But P/Z(P) is isomorphic to a direct summand of P by the splitting hypothesis. Since 0 is the only divisible submodule of P, then Z(P) = P. Hence $x = (1 + M_1, 1 + M_2, 1 + M_3, \ldots) \in Z(P)$. Thus $(0{:}x) = \cap_{i=1}^{\infty}(0{:}1 + M_i) = \cap_{i=1}^{\infty} M_i$ is essential and two-sided in R. By Lemma 1.1 R/(0:x) is a right perfect ring, and hence R/(0:x) has only finitely many maximal ideals. Consequently, R has only finitely many maximal ideals containing (0:x), which contradicts $M_i \supseteq (0{:}x)$ for each i = 1, 2, ... ■

Next we prove two lemmas which are used to prove the main results of this section.

Lemma 1.3. If R is a splitting ring and if every two-sided ideal of R is finitely generated as a left ideal of R, then R/I is a left Artinian ring for any two-sided ideal I which is essential as a right ideal of R.

Proof. By Lemma 1.1 R/I is a right perfect ring. Since each two-sided ideal of R is finitely generated as a left ideal, it follows from an examination of the left Loewy (socle) series of R/I that R/I is a left Noetherian ring. But a left Noetherian, right perfect ring must be left Artinian. ■

Lemma 1.4. Let R be a prime splitting ring with zero right socle such that every two-sided ideal of R is finitely generated as a left ideal of R. Then every two-sided ideal of R has a stationary power.

Proof. Let I be a nonzero two-sided ideal of R. Suppose that $I \supsetneq I^2 \supsetneq I^3 \supsetneq \ldots$ is an infinite descending chain.

Define P_n to be the direct product of $|I^n|$ copies of R/I^{2n} for each $n = 1,2,3, \ldots$. Let $M = \Pi_{n=1}^{\infty} P_n$. Let $p_n = (r+I^{2n})$; i.e., p_n is the element of P_n which has $r + I^{2n}$ at the rth coordinate for each $r \in I^n$. Then $x = (p_1,p_2,p_3,\ldots) \in M$. Since $(0:p_n) = \bigcap_{r \in I^n}(I^{2n}:r)$ is a two sided ideal of R, then $(0:x) = \bigcap_{n=1}^{\infty}(0:p_n)$ is also a two-sided ideal of R. If (0:x) were not 0, then R/(0:x) would be a left Artinian ring by Lemma 1.3. But this contradicts the fact that there is an infinite descending chain

$$(0:p_1) \supsetneq (0:p_1) \cap (0:p_2) \supsetneq \ldots \supsetneq \cap_{i=1}^{n}(0:p_i) \supsetneq \ldots \supsetneq (0:x),$$

as $I^n \subseteq \bigcap_{i=1}^{n}(0:p_i)$ but I^n is not contained in $\bigcap_{i=1}^{n+1}(0:p_i)$. Hence $(0:x) = 0$, so that $x \notin Z(M)$.

An element y of M has infinite I-height in M if $y \in MI^n$ for each positive integer n. Since $P_n I^{2n} = 0$, M has no nonzero elements of infinite I-height.

But $P_n I^{2n} = 0$ implies that any finite sum of the P_n's is contained in Z(M). So, since I is finitely generated as a left ideal,

$x + Z(M)$ has infinite I-height in $M/Z(M)$. But $M/Z(M)$ is a direct summand of M by the splitting hypothesis; so we obtain the desired contradiction (as the image of $x + Z(M)$ in the nonsingular summand of M has infinite I-height in M). ∎

We now come to the two main results of this section, which (along with Lemma 1.3) show that a splitting ring with zero right socle such that every two-sided ideal is finitely generated as a left ideal and a HNP splitting ring have a similar two-sided ideal structure.

Theorem 1.5. Let R be a prime splitting ring with zero right socle such that every two-sided ideal of R is finitely generated as a left ideal. Then R has a unique minimal nonzero two-sided ideal.

Proof. We may assume that R is not simple. By Proposition 1.2, R has only finitely many maximal two-sided ideals. Let I be the intersection of these maximal two-sided ideals. By Lemma 1.4, I has a stationary power, say I^n. Suppose that there exists a nonzero two-sided ideal K which does not contain I^n. Then we may assume (by taking the intersection if necessary) that $K \subsetneq I^n$. By Lemma 1.3 R/K is left Artinian. Hence rad $R/K = I/K$ is nilpotent. Thus there exists an integer m such that $K \supseteq (I^n)^m$. But I^n is idempotent; so $K \supseteq (I^n)^m = I^n \supsetneq K$, which is a contradiction. ∎

If S is a simple module, we use $\mathcal{T}_S$ to denote the smallest torsion class containing S. (See [13]). The S-Loewy (or S-socle) length of a module in $\mathcal{T}_S$ is the (ordinal) number of terms in its ascending S-Loewy (S-socle) series.

Theorem 1.6. Let R be a prime splitting ring with zero right socle such that every two-sided ideal of R is finitely generated as a left ideal. Then every maximal two-sided ideal of R is idempotent.

Proof. First, we show that $\mathrm{Ext}_R(S,S) = 0$ for any simple module S annihilated by a maximal two-sided ideal M. For if $\mathrm{Ext}_R(S,S) \neq 0$, then the S-Loewy (S-socle) length of $E(S) \geq 2$. Any module in $\mathcal{T}_S$ is also in the class of all modules annihilated by the minimal two-sided ideal I of R, which exists by Theorem 1.5. Since R/I must be left Artinian by Lemma 1.3, it follows that the right Loewy (socle) length of R/I is finite. Hence any module in $\mathcal{T}_S$ must also have finite Loewy (socle) length. In particular, the S-Loewy (S-socle) length of $\mathcal{T}_S(E(S))$ is finite, but ≥ 2. Hence there exists a submodule $A \subseteq E(S)$ such that $S \subseteq A$ and $\mathcal{T}_S(E(S)/A) \simeq S$ (as $\mathcal{T}_S$ is closed under extensions). But $E(S)/A$ is injective [13, Corollary 1.5]; hence E(S) is isomorphic to a submodule of E(S)/A. So the S-Loewy (S-socle) length of E(S)/A is ≥ 2, which contradicts $\mathcal{T}_S(E(S)/A) \cong S$. Therefore, $\mathrm{Ext}_R(S,S) = 0$.

By Lemma 1.4 the maximal two-sided ideal M has a stationary power, say M^n. Let $(xR + M^n)/M^n$ be a simple submodule of R/M^n, and choose $K_R \supseteq M^n$ such that K/M^n is maximal with respect to $[K/M^n] \cap [xR + M^n)/M^n] = 0$. Then $R/K \in \mathcal{T}_S$ is an essential extension of $(xR + M^n)/M^n \cong S$. Since $\mathrm{Ext}_R(S,S) = 0$, then $R/M^n = [(xR + M^n)/M^n] \oplus [K/M^n]$. Hence $S \cong (xR + M^n)/M^n$ is a projective R/M^n-module. Since S is the only simple R/M^n-module, it follows that R/M^n is a semi-simple Artinian ring; whence $n = 1$. ■

2. Subidealizers.

In this section we continue the development of the theory of subidealizers, which was begun in [7] and which generalizes the theory of idealizers. (See Introduction for the definitions).

We have been using (0:X) to denote the right annihilator of a set X. Since this section also involves left annihilators, we shall write $\ell(X)$ for the left annihilator of X.

Lemma 2.1. Let M be a right ideal of a semiprime ring T, and let $\ell(M) = 0$ Let R be a subidealizer of M in T. Then the following statements hold:

(1) $(0:M) = 0$;

(2) R is a semiprime ring.

(3) ${}_RM$ is essential in ${}_RT$, and hence ${}_RR$ is essential in ${}_RT$;

(4) If T is right finite dimensional, so is R;

(5) If T is left finite dimensional, so is R;

(6) If T is a prime ring, so is R;

(7) If T has acc on left (right) annihilators, so does R.

Proof. (1) Since $((0:M)M)^2 = 0$ and T is semiprime, then $(0:M)M = 0$; so $(0:M) \subseteq \ell(M) = 0$.

(2) If $A \subseteq R$ and $A^2 = 0$, then $(AM)^2 = 0$; so $AM = 0$ by T semiprime. Since $\ell(M) = 0$, $A = 0$, and hence R is semiprime.

(3) If ${}_RK \subseteq T$ and ${}_RK \cap M = 0$, then $MK \subseteq K \cap M = 0$. Since $(0:M) = 0$ by (1), $K = 0$; so ${}_RM$ is an essential left R-submodule of T.

(4) Let $\sum_{i\in I} A_i$ be a direct sum of nonzero right ideals in R. Since $\ell(M) = 0$, then $\sum_{i\in I} A_i M$ is a direct sum of nonzero right ideals in T. Since T is right finite dimensional, I must be a finite set; so R is also right finite dimensional.

(5) Let $\sum_{i\in I} A_i$ be a direct sum of nonzero left ideals in R. Suppose $t \in (\sum_{i\in I-\{j\}} TA_i) \cap TA_j$. Then

$$Mt \subseteq M(\sum_{i\in I-\{j\}} TA_i) \cap MA_j \subseteq (\sum_{i\in I-\{j\}} A_i) \cap A_j = 0.$$

Since $(0:M) = 0$ by (1), then $t = 0$; so $\sum_{i\in I} TA_i$ is direct. Since T is left finite dimensional, I is a finite set, and hence R is also left finite dimensional.

(6) If A and B are ideals of R such that $AB = 0$, then $(AM)(BM)=0$. Since T is a prime ring, $AM = 0$ or $BM = 0$. Since $\ell(M) = 0$, then $A = 0$ or $B = 0$; so R is a prime ring.

(7) If $A_1 \subseteq A_2 \subseteq A_3 \subseteq \ldots$ is an ascending chain of left (right) annihilators of the subsets $S_1, S_2, S_3, \ldots$, of R, respectively, then we may assume that $S_1 \supseteq S_2 \supseteq S_3 \supseteq \ldots$. By hypothesis the left (right) annihilators $B_1, B_2, B_3, \ldots$ of $S_1, S_2, S_3, \ldots$ in T have acc, and consequently so do $A_1 = B_1 \cap R$, $A_2 = B_2 \cap R$, $A_3 = B_3 \cap R, \ldots$. ∎

The following theorem is an immediate consequence of Lemma 2.1. Moreover, Theorem 2.2 and its following remark give a generalization of [1, Theorems 2.1 and 2.2].

Theorem 2.2. Let T be a ring, and let M be a right ideal of T with $\ell(M) = 0$. Let R be a subidealizer of M in T. If T is a left (right) [semi]prime Goldie ring, then R is also a left (right) [semi]prime Goldie ring.

Remarks. (a) If $\ell(M) = 0$ and T has semisimple left classical quotient ring Q, then Q is also the left classical quotient ring of R. (One can use Lemma 2.1 (1) and [15, Lemma 5] to show that any regular element of R is also regular in T. Using 2.1 (1) and T semiprime left Goldie, one can show that ${}_RR$ is an essential submodule of ${}_RQ$; so it follows by using Theorem 2.2 that every element of Q has the form $d^{-1}r$ for some r, d ε R with d regular in R.)

(b) If $\ell(M) = 0$, if M is an essential right ideal of T, and if T has semisimple right classical quotient ring Q, then Q is also the left classical quotient ring of R. (The proof is similar to (a) except that Lemma 2.5 below must be used to get that R_R is essential in Q_R and that regular elements of R have $0 = (0:x)$ in T.) ■

Lemma 2.3. Let M be a right ideal of a ring T, and let $(0:M) = 0$. Let R be a subidealizer of M in T. Then the following statements hold:

(1) If $\ell(M) = 0$ and $Z(R_R) = 0$, then $Z(T_T) = 0$;

(2) If R is semiprime and $Z({}_RR) = 0$, then $Z({}_TT) = 0$;

(3) If R is semiprime and R is left finite dimensional, then T is left finite dimensional.

Proof. (1) Since $MZ(T_T) \subseteq M \cap Z(T_T)$ and $(0:M) = 0$, it is sufficient to show that $M \cap Z(T_T) = 0$.

Let $x \in M \cap Z(T_T)$, and let K be a nonzero right ideal of R. Since $\ell(M) = 0$, KM is a nonzero right ideal of T, and hence $0 \neq (0:x) \cap KM \subseteq (0:x) \cap K$. Hence $(0:x) \cap R$ is essential in R; so $x \in Z(R_R) = 0$.

(2) First, we show that, for any essential left ideal A of T, $A \cap M$ is an essential left ideal of R. Let L be a nonzero left ideal of R. Then $I = TL \cap A \neq 0$, and $0 \neq MI \subseteq MTL \subseteq L$. Hence $0 \neq MI \subseteq (A \cap M) \cap L$; so $A \cap M$ is an essential left ideal of R.

Let $x \in M \cap Z({}_TT)$. From the preceeding paragraph, $\ell(x) \cap M$ is an essential left ideal of R, and hence $x \in Z({}_RR) = 0$. Thus $MZ({}_TT) \subseteq M \cap Z({}_TT) = 0$. Since $(0:M) = 0$, then $Z({}_TT) = 0$.

(3) Let $\sum_{i\in I} A_i$ be a direct sum of nonzero left ideals in T. Since $(0:M) = 0$, then $\sum_{i\in I} MA_i$ is a direct sum of nonzero left ideals in R. Since R is left finite dimensional, it follows that T must also be left finite dimensional. ■

Our next result generalizes [1, Theorem 1.7] and part of [1, Theorem 1.8].

Theorem 2.4. Let M be a right ideal of a ring T, and let R be a subidealizer of M in T. Then the following statements hold:

(1) If $(0:M) = 0$, $\ell(M) = 0$, and R has a von Neumann regular maximal right quotient ring, then T also has a von Neumann regular maximal right quotient ring;

(2) If $(0:M) = 0$ and R is a semiprime left Goldie ring, then T has a semisimple maximal left quotient ring;

(3) If T is semiprime, $\ell(M) = 0$, and R is a left Goldie ring, then T is also a left Goldie ring.

Proof. (1) and (2) are immediate consequences of Lemma 2.3 and [9, page 106]. (3) follows from Lemma 2.1 (1), (2) and Lemma 2.3. ■

In studying prime splitting rings with zero right socle, we are interested in the case where M is an essential right ideal of T. Hence we shall sharpen our results for the special case where M is essential in T.

Lemma 2.5. Let M be an essential right ideal of a ring T, and let R be a subidealizer of M in T. If R is a semiprime ring, then the following statements hold:

(1) T is a semiprime ring;

(2) M_R is an essential R-submodule of T_R, and hence R_R is essential in T_R;

(3) $\ell(M) = 0$;

(4) If R is right finite dimensional, then so is T;

(5) If R is a prime ring, so is T;

(6) If R has zero right socle, so does T.

Proof. (1) Let A be an ideal of T, and assume $A^2 = 0$. Then $(A \cap M)^2 = 0$. Since R is semiprime and M is essential in T, then A = 0. Hence T is a semiprime ring.

(2) Let K be a right R-submodule of T, and let $M \cap K = 0$. Then $KM \cap M = 0$. Since M_T is an essential right ideal of T, then KM = 0. Thus $(MK)^2 = 0$. Since R is semiprime, MK = 0. Hence $(M \cap KT)^2 = \subseteq MKT = 0$. Since R is semiprime and M_T is essential in T, KT = 0. Thus K = 0, so that M_R is essential in T_R.

(3) Let $x \in \ell(M)$. Then $(xR \cap M)^2 \subseteq xRM \subseteq xM = 0$. Since R is semiprime, $xR \cap M = 0$. Thus by (2), xR = 0; so $\ell(M) = 0$.

(4) If $\sum_{i \in I} A_i$ is a direct sum of nonzero right ideals in T, then $\sum_{i \in I}(A_i \cap M)$ is a direct sum of nonzero right ideals in R (by M essential in T). Thus R right finite dimensional forces T to be right finite dimensional.

(5) If A and B are ideals of T such that AB = 0, then $(A \cap M)(B \cap M) = 0$. Since R is a prime ring, $A \cap M = 0$ or $B \cap M = 0$. But M is essential in T; so A = 0 or B = 0. Hence T is a prime ring.

(6) Let S be a simple T-submodule of T. Since M is an essential right ideal of T, $S \subseteq M$. Since R is semiprime and M is a two-sided ideal of R, xM = S for any nonzero $x \in S$. Thus xR = S for any nonzero $x \in S$; so S is a simple R-submodule of R. ∎

We note that if T is a semiprime right Goldie ring, then $\ell(M) = 0$ for any essential right ideal M of T. Hence the following theorem is a straight forward consequence of Lemmas 2.1, 2.3, and 2.5.

Theorem 2.6. Let M be an essential right ideal of a ring T, and let R be a subidealizer of M in T. Then R is a (semi)prime right Goldie ring if and only if T is a (semi)prime right Goldie ring. R is a (semi)prime left Goldie ring if and only if T is a (semi)prime left Goldie ring and $\ell(M) = 0$.

The following corollary will be of interest in the proof of Theorem 3.4.

Corollary 2.7. Let M be an essential right ideal of a ring T, and let R be a subidealizer of M in T. Then R is a (semi)prime left and right Goldie ring if and only if T is a (semi)prime left and right Goldie ring.

Let M be an essential right ideal of a right nonsingular ring T, and let R be a subidealizer of M in T. From [7, Proposition 4.1] it follows that $T \subseteq Q$, where Q is the maximal right quotient ring of T. If A is a nonsingular R-module, then E(A) is a Q-module [3, 12], and hence a T-module. Thus AT is defined and is a submodule of E(A). An R-module K is called a <u>test module</u> if there exists a nonsingular R-module A (and hence AT) and an exact sequence

$$0 \to K \to \oplus T \to AT/A \to 0.$$

In [7] a right ideal M of a ring T is called <u>generative</u> if TM = T. In the rest of this section, we shall study the relationship of the splitting properties of a ring T and a subidealizer of an essential generative right ideal M of T.

Lemma 2.8. Let T be a splitting ring. Let M be an essential, generative right ideal of T, and let R be a subidealizer of M in T. Let A be a nonsingular R-module, and let K be a test module corresponding to A.

For any singular T-module C, $\mathrm{Ext}_R{}^1(A,C) = 0$ if and only if $\mathrm{Ext}_T{}^1(Z(K \otimes T),C) = 0$.

Proof. Consider the following commutative exact diagram:

$$\begin{array}{ccccccc}
\mathrm{Hom}_R(AM,C) & \xrightarrow{\beta} & \mathrm{Ext}_R(AT/AM,C) & \to & \mathrm{Ext}_R(AT,C) & & \\
\alpha\downarrow & & \gamma\downarrow & & & & \\
\mathrm{Hom}_R(AM,C) & \xrightarrow{\delta} & \mathrm{Ext}_R(A/AM,C) & \to & \mathrm{Ext}_R(A,C) & \to & \mathrm{Ext}_R(AM,C) \\
 & & \downarrow & & & & \\
 & & \mathrm{Ext}_R^2(AT/A,C) & & & & \\
 & & \downarrow & & & & \\
 & & \mathrm{Ext}_R^2(AT/AM,C) & & & &
\end{array}$$

Since T is a splitting ring, then by [7, Proposition 1.2] we have $\mathrm{Ext}_R(AM,C) \cong \mathrm{Ext}_T(AM,C) = 0$ and $\mathrm{Ext}_R(AT,C) \cong \mathrm{Ext}_T(AT,C) = 0$. Thus β is an epimorphism. By [13, Theorem 1.3], inj dim $C_T \leq 1$; so $\mathrm{Ext}_R{}^2(AT/AM,C) \cong \mathrm{Ext}_T{}^2(AT/AM,C) = 0$ by [7, Proposition 1.2]. Consequently, it is easy to verify from the diagram that each of the following statements is equivalent to the next:

(1) $\mathrm{Ext}_R(A,C) = 0$;

(2) δ is an epimorphism;

(3) $\gamma\beta$ is an epimorphism;

(4) γ is an epimorphism;

(5) $\mathrm{Ext}_R{}^2(AT/A,C) = 0$.

Thus it is sufficient to show that $\mathrm{Ext}_R{}^2(AT/A,C) \cong \mathrm{Ext}_T(Z(K\otimes T),C)$, The following exact sequence is exact: $\mathrm{Ext}_R(\oplus T,C) \to \mathrm{Ext}_R(K,C) \to \mathrm{Ext}_R{}^2(AT/A,C) \to \mathrm{Ext}_R{}^2(\oplus T,C)$. But $\mathrm{Ext}_R(\oplus T,C) = 0 = \mathrm{Ext}_R{}^2(\oplus T,C)$ by [7, Proposition 1.2]; so it follows [7, Proposition 5.1] and the splitting ring hypothesis that $\mathrm{Ext}_R{}^2(AT/A,C) \cong \mathrm{Ext}_R(K,C) \cong \mathrm{Ext}_T(K \otimes T,C) \cong \mathrm{Ext}_T(Z(K \otimes T),C)$. ■

Let M be a right ideal of a ring T, and let R be a subidealizer of M in T. R is a <u>tame</u> subidealizer of M in T if R/M is a semisimple Artinian ring [7]. R is a <u>very tame</u> subidealizer of M in T if R/M is a simple Artinian ring.

Theorem 2.9. Let M be an essential, generative right ideal of the right nonsingular ring T, and let R be a tame subidealizer of M in T. Then R is a splitting ring if and only if T is a splitting ring and $\mathrm{Ext}_T{}^1(Z(K_R \otimes T),C) = 0$ for all singular C_T and all test modules K.

Proof. "Only if": If R has SP, then T has SP by [7, Proposition 4.1 and 1.2]. C_T is a singular R-mod by [7, Proposition 4.1]. Since R has SP, then $\mathrm{Ext}_R{}^1(A,C) = 0$ for all nonsingular A_R and singular C_T. Hence $\mathrm{Ext}_T{}^1(Z(K \otimes T),C) = 0$ by Lemma 2.8.

"If": We must show $\mathrm{Ext}_R{}^1(A,C) = 0$ for all nonsingular A_R and all singular C_R. It is sufficient to show $\mathrm{Ext}_R{}^1(A,C/CM) = 0$ and $\mathrm{Ext}_R{}^1(A,CM) = 0$. Since $M^2 = M$, we may assume WLOG that $CM = 0$ or $CM = C$.

Case 1: $CM = 0$. We first show that $\mathrm{Tor}_R(A,R/M) = 0$. Since T_R is projective and $TM = T$, then $T \otimes_R M \to T \otimes_R R \to T$ is an isomorphism. Since A is nonsingular, AT is defined; so $AT \otimes_R M \cong AT \otimes_T T \otimes_R M \cong AT \otimes_T T \cong AT$. Since ${}_RM$ is projective [7, Proposition 1.1], $\mathrm{Tor}_1{}^R(AT/A,M) = 0$. Hence $A \otimes_R M \to AT \otimes_R M \to AT$ is mono, and thus $A \otimes_R M \to A \otimes_R R$ is also mono. Therefore $\mathrm{Tor}_1{}^R(A,R/M) = 0$.

Now consider the sequence

$$E:\quad 0 \to C \to B \to A \to 0.$$

Since $\mathrm{Tor}_1{}^R(A,R/M) = 0$, we obtain the induced sequence

$$E^*:\quad 0 \to C \to B/BM \to A/AM \to 0.$$

Since R/M is semisimple Artinian, E^* splits, and hence E splits.

Case 2: $CM = C$. Here $C \cong P/J$ for a direct sum P of copies of M. It suffices to show $\mathrm{Ext}_R{}^1(A,P/JM) = 0$ and $\mathrm{Ext}_R{}^2(A,J/JM) = 0$.

Note that P/JM is a singular T-module [7, Proposition 4.1]; so the hypothesis and Lemma 2.8 imply $\mathrm{Ext}_R{}^1(A,P/JM) = 0$ as desired.

Next choose an exact sequence $0 \to L \to F \to A \to 0$ with F_R free.

Since L is nonsingular, then $\mathrm{Ext}_R{}^2(A,J/JM) \cong \mathrm{Ext}_R{}^1(L,J/JM) = 0$ by case 1. ■

As corollaries to Theorem 2.9, we obtain the following two theorems of Goodearl.

Corollary 2.10 [6, Theorem 10]. Let M be a finite intersection of essential maximal right ideals of the right nonsingular ring T, and let R be the idealizer of M in T. Then R is a splitting ring if and only if T is a splitting ring.

Proof. By [10, Proposition 1.7 and 1.1], we may assume that M is generative and R is tame. By [6, Proposition 9], $Z(K \otimes T) = 0$ for each test module K; so the result follows from Theorem 2.9. ■

Corollary 2.11 [7, Theorem 5.2]. Let M be an essential, generative right ideal of the right nonsingular ring T, and let R be a tame subidealizer of M in T. If T is a splitting ring and if $\mathrm{Ext}_T(T/M,C) = 0$ for all singular C_T, then R is a splitting ring.

Proof. By [7, Lemma 5.1 (b)], $Z(K \otimes_R T) \subseteq \oplus T/M$ for any test module K. Thus we have an exact sequence for each singular C_T: $\mathrm{Ext}_R{}^1(\oplus T/M,C) \to \mathrm{Ext}_R{}^1(Z(K \otimes T),C) \to \mathrm{Ext}_R{}^2(\frac{\oplus T/M}{Z(K \otimes T)}, C)$, where the left end is 0 by hypothesis. By [7, Proposition 1.2 and 13, Theorem 1.3], $\mathrm{Ext}_R{}^2(\frac{\oplus T/M}{Z(K \otimes_R T)}, C) \cong \mathrm{Ext}_T{}^2(\frac{\oplus T/M}{Z(K \otimes_R T)}, C) = 0$. Hence $\mathrm{Ext}_R{}^1(Z(K \otimes T),C) = 0$; so the result follows from Theorem 2.9. ■

A subidealizer R of a right ideal M in a ring T is called proper if R is not equal to the idealizer of M in T.

Proposition 2.12. Let M be an essential generative right ideal in a right nonsingular ring T, and let R be a very tame proper subidealizer of M in T. Then R is splitting ring if and only if T is splitting ring and $\mathrm{Ext}_T(T/M,C) = 0$ for all singular C_T.

Proof. The "if" part follows from Corollary 2.11.

"Only if". Since R is right splitting, then T is right splitting by [7, Proposition 4.1 and 1.2].

Let S be the idealizer of M in T. Since R is a proper subidealizer $S \supsetneq R$. Since $H = \{x \in R \mid (S/R)x = 0\}$ is a two sided ideal of R containing M and since $S \neq R$, then $H = M$ by the maximality of M. Hence S/R is a direct sum of simple modules annihilated by M. Hence R/M is a direct summand of a direct sum of copies of S/R.

By Goodearl [7, Lemma 2.4(a)], $\mathrm{Tor}_1^R(S/R,T) = 0$; hence we obtain the exact sequence

$$0 \to R \otimes_R T \to S \otimes_R T \to S/R \otimes_R T \to 0.$$

The composition $R \otimes_R T \to S \otimes_R T \to T \to R \otimes_R T$ is the identity map on $R \otimes_R T$. Hence the exact sequence splits. Thus $[(S/R) \otimes_R T]_T$ is a direct summand of $[S \otimes_R T]_T$.

Now let C_T be a singular right T-module. Then $\mathrm{Ext}_R^1(S,C) = 0$ by R splitting. By [7, Proposition 5.1 (d)], $\mathrm{Ext}_T^1(S \otimes T,C) = 0$. By the preceeding paragraph, $\mathrm{Ext}_T^1((S/R) \otimes T,C) = 0$. Since R/M is a direct summand of a direct sum of copies of S/R, then $T/M = T/MT \cong R/M \otimes_R T$ is a direct summand of a direct sum of copies of $S/R \otimes_R T$. Hence $\mathrm{Ext}_T^1(T/M,C) = 0$. ∎

§3. Embedding Theorems.

In this section we show that prime splitting rings with zero right socle and enough finiteness conditions are iterated very tame subidealizers of simple splitting rings with finiteness conditions. A considerable amount is known about simple hereditary splitting rings [8]; no example of simple nonhereditary splitting ring is known. (Any splitting ring has homological dimension ≤ 2 [13, Theorem 2.2].)

Before we state our embedding theorems, we need to review some terminology and prove several lemmas.

A ring R is called an <u>iterated</u> (<u>very</u> <u>tame</u>) <u>subidealizer</u> of a ring T is there exists a chain $R = R_0 \subset R_1 \subset R_2 \subset \ldots \subset R_n = T$ of rings such that each R_i ($i = 0, 1, 2, \ldots, n-1$) is a (very tame) subidealizer of an essential, generative right ideal of R_{i+1}.

A hereditary torsion class [13] is called TTF if it is closed under products. A hereditary torsion class if TTF if and only if its associated filter of right ideals has a minimal member. For an explanation of perfect torsion theories as well as an exposition on quotient rings and relative injectivity associated with a hereditary torsion class, the reader is referred to [12].

Lemma 3.1. Let M be an essential, generative right ideal of a ring T, and let R be a tame subidealizer of M in T. Then the following statements hold for the class $\mathfrak{T}_M$ of left R-modules annihilated by M.

(1) $\mathfrak{T}_M$ is a perfect torsion class whose quotient ring is T.

(2) If $\mathfrak{T}_M{}^*$ is the torsion functor associated with $\mathfrak{T}_M$, then $\mathfrak{T}_M{}^*(T \otimes_R A) = 0$ for every left R-module A.

(3) If $\mathfrak{T}_M{}^*(A) = 0$, then the multiplication map $T \otimes_R A \to TA$ is an isomorphism, and the canonical map $A \to T \otimes_R A$ is a monomorphism.

Proof. (1) Since M is generative, $M^2 = M$; so $\mathfrak{T}_M$ is a TTF class. By [7, Proposition 1.1], M is a finitely generated, projective left R-module; so $\mathfrak{T}_M$ must be perfect. Since TM = T, T is the quotient ring.

(2) follows from [12, Ex. 1, p. 81] and (1).

(3) follows from [12, Theorem 13.1] and (1). ■

Lemma 3.2. Let M be an essential generative right ideal of T, and let R be a tame idealizer of M in T. Let $\mathfrak{X}$ denote the set of maximal left ideals of R which do not contain M.

(1) For each $N \in \mathfrak{X}$, TN is a maximal left ideal of T, and the

natural map $f: R/N \to T/TN$ is an essential monomorphism of left R-modules.

(2) For any $K, N \in \mathfrak{X}$, $T/TK \cong T/TN$ if and only if $R/K \cong R/N$.

(3) If R is a prime splitting ring, then any bounded simple left T-module A is isomorphic to T/TN for some $N \in \mathfrak{X}$.

(4) If R is a splitting ring, then T is also a splitting ring.

(5) If R is a prime splitting ring, then there is a monic correspondence from the set of maximal two-sided ideals of T to the set of maximal two-sided ideals of R which do not contain M.

(6) If every two-sided ideal of R is finitely presented as a left ideal, then every two-sided ideal of T is finitely presented as a left ideal.

Proof. (1) Note that TN/N is annihilated by M, while R/N is not; hence $(R/N) \cap (TN/N) = 0$, from which it follows that f is a monomorphism. Given any $x \in T - TN$, we have $Mx \subseteq M \subseteq R$. Since $T/TN \cong T \otimes_R R/N$ has no submodule annihilated by M by Lemma 3.1(2), then $Mx \nsubseteq TN$. Hence $Mx \nsubseteq N$; so $Rx \cap R \nsubseteq N$. Therefore, $f(R/N)$ is essential in T/TN. Now T/TN is a simple T-module by [11, Corollary 1.5]; so TN must be a maximal left ideal of T.

(2) If $R/N \cong R/K$, then

$$T/TN \cong T \otimes_R (R/N) \cong T \otimes_R (R/K) \cong T/TK.$$

Conversely, by (1), T/TN has R/N as an essential simple submodule, and T/TK has R/K as an essential simple submodule. So the isomorphism $T/TN \cong T/TK$ must carry R/N onto R/K. Thus $R/N \cong R/K$.

(3) Since A is a bounded, simple, left T-module, it is annihilated by a nonzero two-sided ideal K of T. Then $K \cap R$ is a two-sided ideal of R which annihilates A. Hence A is an $R/(R \cap K)$-module. Since R is prime $K \cap R$ is essential in R by Lemma 2.5(2). Hence $R/(R \cap K)$ must be right perfect by Lemma 1.1. Hence A has nonzero socle as a left $R/(R \cap K)$-module. Since $T \otimes_R A \cong {}_TA$, no simple module in the socle of ${}_RA$ is annihilated by M in Lemma 3.1 (2). Thus a simple submodule of

of ${}_R A$ is isomorphic to R/N for some $N \in \mathfrak{X}$. By (1) and the flatness of T_R, we have $0 \neq T/TN \cong T \otimes_R R/N \subseteq T \otimes_R A \cong T \otimes_T A \cong A$. Since ${}_T A$ is simple, then $T/TN \cong A$.

(4) This follows from [7, Proposition 4.1 and 1.2].

(5) Let N' be a maximal two-sided ideal of T. It follows from (4), Lemma 2.5 (5), and Lemma 1.1 that T/N' is a simple Artinian ring. Let A be a simple left T-submodule of T/N'. By (3) $A \cong T/TN$ for some $N \in \mathfrak{X}$. Since $N'A = 0$, then $(N' \cap R)R/N = 0$ by (1).

Thus N contains a maximal two-sided ideal of R, namely $N" = \cap \{N_\alpha | R/N \cong R/N\}$. (We note that this is a finite intersection, since R is a splitting ring and hence $R/N"$ is right perfect.) Note that $N" \neq M$ since $N \in \mathfrak{X}$.

Thus we have a correspondence $N' \to N"$. By (2) the correspondence must be well-defined. Moreover, the correspondence is monic. For let K' be a maximal ideal of T distinct from N', and let B be the simple R-submodule which can be embedded in R/K'. Since R/K' is right perfect by Lemma 1.1, then by (3) $T/K' \cong \oplus T/TK$ for some $K \in \mathfrak{X}$. Since $K' \neq N'$, $T/TN \ncong T/TK$; so $R/N \ncong R/K$ by (2). Consequently, $K" = \cap \{K_\alpha | R/K_\alpha \cong R/K\} \neq \cap \{N_\alpha | R/N_\alpha \cong R/N\} = N"$.

(6) Let $K"$ be a two-sided ideal of T, and let $K = R \cap K'$. Then there is a natural monomorphism $g: R/K \to T/K'$. Since T_R is flat, $1 \otimes g: T \otimes_R (R/K \to T \otimes_R (T/K')$ is a monomorphism. By [7] $T \otimes_R (T/K') \cong T \otimes_T (T/K') \cong T/K'$. So under the composition of these monomorphisms, $(k' \otimes (r + k)) \to k'r + K' = K'$ for any $k' \in K'$. Hence $\{\sum(k' \otimes (r + K)) | k' \in K', r \in R\} \subseteq \ker(1 \otimes g) = 0$. But $T \otimes_R (R/K) \cong T/TK$, and under this isomorphism $\{\sum(k' \otimes (r + K)) | k' \in K', r \in R\}$ maps onto K'/TK. Hence $TK = K'$.

Now since ${}_R K$ is finitely presented, there exists an exact sequence

$$0 \to {}_R L \to {}_R F \to {}_R K \to 0,$$

where ${}_R L$, ${}_R F$ are finitely generated and ${}_R F$ is free. Since T_R is flat, then

$$0 \to T \otimes_R L \to T \otimes_R F \to T \otimes_R K \to 0,$$

is an exact sequence of finitely generated left T-modules, with $T \otimes_R F$ a free left T-module. By Lemma 3.1 (3), $T \otimes_R K \cong TK = K'$; so ${}_TK'$ is finitely presented. ∎

Lemma 3.3. Let R be a prime, left and right Goldie, splitting ring. Suppose that M is a nonzero, idempotent, maximal two-sided ideal of R such that M is finitely presented as a left ideal of R. Let Q be the simple classical quotient ring of R, and let $T = \{q \in Q | M_q \subseteq R\}$. Then the following statements hold:

(1) T is the quotient ring for the perfect TTF class $\mathcal{T}_M$ of left R-modules annihilated by M.

(2) M is an essential, generative right ideal of T, and T is a very tame subidealizer of M in T.

Proof. (1) Since $M^2 = M$ and ${}_RQ \cong E(R)$, then T must be the quotient ring for $\mathcal{T}_M$. Since R is a splitting ring, ${}_RM$ is flat; so, since ${}_RM$ is finitely presented, ${}_RM$ is projective. Thus $\mathcal{T}_M$ must be perfect.

(2) By (1) and [12, Theorem 13.1], TM = T. Also we have the exact sequence

$$0 \to \mathrm{Hom}_R({}_RM, {}_RM) \to \mathrm{Hom}_R({}_RM, {}_RR) \to \mathrm{Hom}_R({}_RM, R/M).$$

Since $M^2 = M$ is two sided, Hom(M,R/M) = 0. Hence $\mathrm{Hom}_R(M,M) \cong \mathrm{Hom}_R(M,R)$. So if $Mq \subseteq R$, then $Mq \subseteq M$; thus M is a right ideal of T. Since R is prime, ${}_RM_R$ is essential in R; so, since $R \subseteq T \subseteq Q$ and R_R is essential in Q_R, then M_R is essential in T_R. Hence M_T is essential in T_T. Since M is a maximal two-sided ideal of R, it follows from Lemma 1.1 that R is a very tame subidealizer of M in T. ∎

We can now state our first embedding theorem.

Theorem 3.4. Let R be a prime, left and right Goldie, splitting ring with zero right socle. Suppose that each two-sided ideal of R is finitely presented as a left ideal. Then R is an iterated very tame subidealizer of a simple, left and right Goldie, splitting ring T with zero right socle.

Proof. By Proposition 1.2, R has only finitely many maximal two-sided ideals; say n.

Since R is a splitting ring, each maximal two-sided ideal is idempotent (by Theorem 1.6) and projective as a left ideal (as in the proof of Lemma 3.3).

Let M be a maximal two-sided ideal of R. Then we can form the quotient ring $T_1 \supseteq R$ for the TTF class $\mathfrak{T}_M$ of all left modules annihilated by the maximal two-sided ideal M. By Lemma 3.3, M is an essential, generative right ideal of T_1 and R is a very tame subidealizer of M in T_1. By Corollary 2.7, T_1 is a prime, left and right Goldie ring. By Lemma 3.2 (4), T_1 is also a right splitting ring. It follows from Lemma 3.2 (5) that T_1 has at most n-1 maximal two-sided ideals. Each of the two-sided idealt of T_1 is finitely presented as a left T_1-module by Lemma 3.2 (6). Finally, T_1 has zero right socle by Lemma 2.5 (6).

Now by repeated applications of the process of the above paragraph, we form subrings T_i (i = 2, 3, ...k) of the classical quotient ring Q of R such that T_{i-1} is a very tame subidealizer of an essential generative right ideal of T_i. Moreover, we can continue this process if and only if T_{i-1} is not simple; so the process continues until we reach a simple ring $T = T_K$. (We note that $k \leq n$ because of Lemma 3.2 (5).) ∎

Remarks. (i) All known examples of prime splitting rings satisfy the hypotheses of Theorem 3.4.

(ii) We also have considerable information about the converse of Theorem 3.4. If T is a simple, left and right Goldie, splitting ring with zero right socle, then by Corollary 2.7 any iterated very tame subidealizer R of T is a prime left and right Goldie ring, and Theorem 2.9 (or its corollaries) gives us necessary and sufficient conditions to determine whether R is a splitting ring. Also at each stage of the iterated idealizer we introduce a two-sided maximal ideal, which is finitely presented left ideal at that stage, by [7, Proposition 1.1(a)].

In [7] there is an example of a prime splitting ring which is neither left nor right Noetherian. However, since many interesting splitting rings are Noetherian, we point out the following corollary to Theorem 3.4.

Corollary 3.5. Let R be a prime, left and right Noetherian, splitting ring with zero right socle. Then R is an iterated very tame subidealizer of a simple, left Goldie, right Noetherian, splitting ring T with zero right socle.

Proof. Everything in the conclusion follows from Theorem 3.4 except the right Noetherian property of T. But, by an iterative use of [7, Proposition 1.1 (a)], it follows that T is a finitely generated right R-module. Since R is a right Noetherian ring, it then follows that T must be right Noetherian also. ∎

In Theorem 3.4 the embedding of certain splitting rings into simple rings was done by an iterative formation of quotient rings in such a way that each was a very tame subidealizer in the next. This process depended on knowing that the two-sided ideals were finitely presented as left ideals. Under slightly weaker finiteness conditions, our next result shows that a "looser" embedding can in fact be accomplished by the formation of only one quotient ring.

Let M be a right ideal of T, and let R be a subidealizer of M in T. R is called (right) <u>controlled</u> if R/M is left Artinian.

Theorem 3.6. Let R be a prime, left and right Goldie, splitting ring with zero right socle. Suppose that every two-sided ideal of R is finitely generated as a left ideal. If the (necessarily unique) minimal nonzero two-sided ideal of R is finitely presented as a left ideal, then there exists a simple, left and right Goldie, splitting ring T with zero right socle such that (a) I is an essential generative right ideal of T, and (b) R is a controlled subidealizer of I in T.

Proof. By Theorem 1.5 R has a unique minimal nonzero two-sided ideal I. We may assume $I \neq R$. By hypothesis ${}_RI$ is finitely related and essential in R. Since ${}_RI$ is flat by [3, Proposition 4.2], then ${}_RI$ is projective. Thus the TTF class of left modules annihilated by I ($I^2 = I$ by Theorem 1.5 and the primeness of R) is a perfect torsion class. The quotient ring associated with this class is

$$T = \{q \in Q \mid Iq \subseteq R\},$$

where Q is the simple classical quotient ring of R. Note that I_R is essential in T_R. As in the proof of Lemma 3.3, TI = T and IT = I; also I_T is essential in T_T. Since R/I is left Artinian by Lemma 1.3, then R is a controlled subidealizer of I in T. ∎

PROBLEM SESSION ON REGULAR RINGS

Chaired by

Joe W. Fisher
The University of Texas at Austin
Austin, Texas

The theory of regular rings springs from the theory of operator algebras. In particular, Von Neumann's study of projection lattices of certain operator algebras led him to introduce continuous geometries (a kind of lattice) and regular rings (which he used to coordinate certain continuous geometries, in a manner analogous to the introduction of division ring coordinates in projective geometry). Regular rings also arise naturally in the study of measure theory--namely, as Boolean algebras. References, in this regard, include I. Halperin [16], I. Kaplansky [21], F. Maeda [24], J. Von Neumann [32], and L. A. Skornyakov [30].

Throughout the years there has been a great deal of interest and research activity in regular rings as witnessed by the literature. Recently, there has been a flourish of activity, not only on regular rings [1, 7, 8, 9], regular self-injective rings [11, 12, 13, 14, 17, 27], unit regular rings [5, 10, 18, 19], regular rings generated by their units [3, 4, 10, 20, 31]; but also, on such related topics as V-rings [1, 6, 7, 25] and Von Neumann or direct finiteness [14, 18, 19, 23, 27]. The lists of references indicated are not complete nor are they intended to be. They are merely to indicate some of the areas of recent activity together with some of the people who are conducting it. The bibliographies of the above references will give the reader a much broader picture. Some of the problems listed are new--others are listed for updating and re-emphasis.

1. In [7] the author gave an expository account of regular rings, V-rings, and their interconnections. Six problems were posed there which will not be reproduced here. At the time of writing, the only progress known to the author was that R. L. Snider has shown that countable π-regular V-rings are regular.

2. In 1958, L. A. Skornyakov [30, p. 167] conjectured that all regular rings are generated by their units. In 1974 G. Bergman [18] settled it in the negative and Fisher-Snider [10] proved the conjecture in the affirmative for a large class of regular rings. Closely tied to 'generation by units' are the concepts of 'unit regularity' (for each x, there exists a unit u, such that $x = xux$) and 'direct finiteness' ($xy = 1$ implies $yx = 1$). Some interesting problems which remain are the following:

(a) [G. Bergman] If R is a directly finite simple regular ring then is R unit regular?

(b) [K. Goodearl] If R is a directly finite regular ring and M is a maximal ideal of R, then is R/M directly finite?

(c) [M. Henriksen[19]] If R is a regular elementary divisor ring, then is R unit regular? If, in addition, R satisfies a polynomial identity, then the answer is affirmative [10].

(d) [Fisher-Snider] If R is regular and each prime factor ring of R is unit regular (generated by its units), then is R unit regular (generated by its units)?

3. An important question is the following: If R is a directly finite regular ring, then is the matrix ring R_n directly finite for each n? Without the regularity, the answer is negative [29]. If in addition, R is self-injective [14, 27], or R is unit regular [19], then the answer is affirmative. If R is a regular Baer *-ring with left projections *-equivalent to right projections, then the answer is affirmative by [2, Exercise 2, p. 257]. More generally, if R is a directly finite Baer *-ring with left projections equivalent to right

projections and with sufficiently many projections, then R_n is directly finite by [2, loc. cit.] and the results in [15, 26] (Thanks to S. K. Berberian for bringing this to my attention).

4. (J.-E. Roos [28]) If R is a directly finite, regular, right self-injective ring, then is R left self-injective? This is asserted in [28, Remàrque 2]; however, Roos has admitted that his proof is incomplete. Jérémy and Gousaud have made some progress on this one when, in addition, R is upper continuous and lower χ_0-continuous.

6. (K. Goodearl) If R is unit regular with A and B finitely generated projective right R-modules such that $A^{(n)} \cong B^{(n)}$, then is $A \cong B$? Or if $A^{(n)}$ is subisomorphic to $B^{(n)}$, then is A subisomorphic to B? See [17, Problem 4)

7. (W. D. Blair) If R is a prime regular ring which is integral over its center, then is R primitive? This is a special case of a conjecture of Kaplansky [22] that every prime regular ring is primitive. Kaplansky's conjecture has been settled in the cases where R is self-injective [11] or R is countable [8].

8. (R. L. Snider) If R is a prime regular ring which is both a left and right V-ring, then is R Artinian?

9. (S. Steinberg) If R is right self-injective, left valuation ring, then is R left self-injective?

10. (S. K. Jain) If R is a prime local left and/or right self-injective ring, then is R a division ring? John Lawrence reportedly has a counterexample but I have not received conformation from him yet!

REFERENCES

1. Armendariz, E. P. and J. W. Fisher, "Regular P. I.-rings," Proc. Amer. Math. Soc. 39(1973), 247-251.

2. Berberian, S. K., Baer *-Rings, Springer-Verlag, Berlin, 1972.

3. Burgess, W. D. and W. Stephenson, "Pierce sheaves and rings generated by their units," (to appear).

4. ________________, "Pierce sheaves of non-commutative rings," (to appear).

5. Ehrlich, G., "Units and one-sided units in regular rings," (to appear).

6. Farkas, D. R. and R. L. Snider, "On group algebras whose simple modules are injective," Trans. Amer. Math. Soc. 194(1974), 241-248.

7. Fisher, J. W., "Von Neumann regular rings versus V-rings, Proc. of University of Oklahoma ring theory symposium, Marcel Dekker, Inc., 101-119.

8. Fisher, J. W. and R. L. Snider, "Prime von Neumann regular rings and primitive group algebras," Proc. Amer. Math. Soc. 44(1974), 244-250.

9. ________________, "On the von Neumann regularity of rings with regular prime factor rings," Pacific J. Math. 53(1974), 138-147.

10. ________________, "Rings generated by their units," (to appear).

11. Goodearl, K. R., "Prime ideals in regular self-injective rings," Canad. J. Math. 25(1973), 829-839.

12. ________, "Prime ideals in regular self-injective rings, II" J. Pure and Appl. Alg., 3(1973), 357-373.

13. Goodearl, K. R., and A. K. Boyle, "Dimension theory for non-singular injective modules," (to appear).

14. Goodearl, K. R., and D. Handelman, "Simple self-injective rings," (to appear).

15. Hafner, I., "The regular ring and the maximal ring of quotients of a finite Baer *-ring," Mich. Math. J., 21(1974), 153 160.

16. Halperin, I., Introduction to von Neumann algebras and continuous geometry, Canad. Math. Bull. 3(1960), 273-288.

17. Handelman, D., "Simple regular rings with a unique rank function," (to appear).

18. ________, "Perspectivity and cancellation in regular rings," (to appear).

19. Henriksen, M., "On a class of regular rings that are elementary divisor rings," Arch. Math., 34(1973), 133-141.

20. ________, "Two classes of rings generated by their units," J. Algebra, 31(1974), 182-193.

21. Kaplansky, I., Rings of Operators, New York: Benjamin 1968.

22. ________, Algebraic and analytic aspects of operator algebras. CBMS Regional Conference Series in Mathematics, No. 1. Providence, R. I., Amer. Math. Soc. 1970.

23. Losey, G., "Are one-sided inverses two-sided inverses in a matrix ring over a group ring?" Canad. M. Bull., 13(1970), 475-479.

24. Maeda, F., Kontinuierliche Geometrien. Berlin-Gottingen-Heidelberg: Springer 1958.

25. Michler, G. and O. Villamayor, "On rings whose simple modules are injective," J. Algebra 25(1973), 185-201.

26. Pyle, E. S., "The regular ring and the maximal ring of quotients of finite Baer *-rings," Trans. Amer. Math. Soc., 203(1975), 201-214.

27. Renault, G., "Anneaux régulier auto-injectifs à droite," Bull. Soc. Math. France, 101(1973), 237-254.

28. Roos, J.-E., "Sur l'anneau maximal de tractions des AW*-algèbres et des anneaux de Baer," C. R. Acad. Sci. Paris 266(1968), 120-123.

29. Shepherdson, J. C., "Inverse and zero divisors in matrix rings," Proc. Lond. Math. Soc. (3)1(1951), 71-85.

30. Skornyakov, L. A., Complemented modular lattices and regular rings. Edinburgh: Oliver and Boyd, 1964.

31. Stephenson, W., "Rings which are generated by their nilpotents, itempotents, or units," (to appear.

32. Von Neumann, J., Continuous Geometry, Princeton (1960) Princeton University Press.

PROBLEMS

Five problem sessions were held during the conference, each of the sessions being taped in its entirety. Most of the problems listed below were posed during these sessions. In order not to further delay publication of the proceedings, Professors Cozzens and Sandomierski transcribed the problems appearing on the tapes, adding comments, etc., when deemed appropriate. If there are any glaring errors or omissions, these are solely the responsibility of the editors and should not be attributed to the indicated proposer. Finally, it is conceivable that some of the problems have been resolved in the interim. To avoid any disappointment, the interested reader should contact the appropriate proposer(s) for a status report on the relevant problem(s).

George Bergman:

1. Let $\mathcal{F}_R$ ($_R\mathcal{F}$) be the category of finitely presented right (left) R-modules having homological dimension ≤ 1 and satisfying $_RM^* = \mathrm{Hom}(M_R, R_R) = 0$ ($M_R^* = 0$). Then $\mathrm{Ext}_R^1(-, R)$ defines a duality between $\mathcal{F}_R$ and $_R\mathcal{F}$. More generally, $\mathrm{Ext}_R^n(-, R)$ defines a duality between the category of all finitely presented right R-modules having homological dimension $\leq n$ which do not map non-trivially into modules of dimension $< n$ and the corresponding class of finitely presented left R-modules.

If M is an arbitrary right R-module is it possible to express M in terms of a series of extensions using modules fitting into each of the above classes. For example, if one considers the exact sequence $0 \to K \to M \to M^{**}$, M^{**} may be projective and K a member of the next class. Consequently, one might be able to reduce questions about general modules to questions about modules in these classes for which we have a duality of the aforementioned type.

2. Let R be a PI-domain, D its quotient field and $f_i : R \to E_i$ a finite family of ring homomorphisms into division rings E_i. Suppose $a \in R$ satisfies $f_i(a) \neq 0$ for each i. If we adjoint $a^{-1} \in D$ to R some of the f_i may not extend to $R[a^{-1}]$. However, a natural question to ask is, when will this process of adjoining inverses lead to an extension R' of R and corresponding extensions of the f_i with the property that each nonzero $a \in R'$ has a zero image in some E_i?

Such an R' will necessarily be semilocal with the E_i's, its residue fields. There are certain restrictions on R and the f_i which are implied by the results in Bergman-Small [75], and there are examples, which are in fact Noetherian, where this process fails to work.

[75] Bergman, G. and Small, Lance, P. I. degrees and prime ideals, J. of Algebra, 33(1975), 435-462.

John H. Cozzens:

1. Let δ be a ρ-derivation of a noncommutative field k with ρ nonsurjective. Can one extend $\delta(\rho)$ to a field $\bar{k} \supset k$ in such a manner that

a. every linear differential equation in δ has a solution in $\bar{k}$,

b. ρ is still nonsurjective.

If this is possible, the ring of ρ-differential operators with coefficients in $\bar{k}$ will be an example of a right Noetherian right V-domain which is not a left V-domain (e.g., see Cozzens-Faith [75], Theorem 5.21).

Conjecture: Yes, in some instances.

[75] Cozzens, J. H. and Faith, C., "Simple Noetherian Rings," Cambridge Tracts in Mathematics, No. 69, Cambridge University Press, Cambridge.

J. H. Cozzens and F. L. Sandomierski:

Throughout R will denote a two-sided Noetherian, prime maximal order and P, a nonzero invertible prime ideal of R.

1. If R is 2-dimensional (i.e., glbR = 2) when is $\bar{R} = R/P$ hereditary (Dedekind prime)?

Remark: inj. dim $_{\bar{R}}\bar{R} = 1$.

2. More generally, when is $\text{glb}\bar{R} = \text{glb}R - 1$?

3. If R is 2-dimensional and P is square free, i.e., whenever $P \subset M$, M maximal, $P \not\subset M^2$, does this imply that $\bar{R}$ is Dedekind prime?

4. When R is 2-dimensional, need R have non-trivial idempotent two-sided ideals? In particular, does a classical 2-dimensional maximal R-order, R a 2-dimensional regular local ring have non-trivial idempotent ideals?

By Riley [72], 2-dimensional maximal quaterion orders have no non-trivial idempotent ideals.

5. Suppose R is 2-dimensional and has no non-trivial reflexive ideals. Does this imply that R is simple?

When R is 1-dimensional, the answer is trivially yes.

For dimension > 2, the answer is no (see Cozzens [76]).

[72] Riley, J. A., "Maximal quaterion orders," J. of Albegra, 3(1972), 241-249.

[76] Cozzens, J. H., "Maximal orders and reflexive modules, Trans. Amer. Math. Soc., 219(1976), 1-14.

Robert Gordon:

1. Does there exist a simple (two-sided) Noetherian ring R with rt kd R = 1 and lkd R   1?

2. More generally, does there exist a prime Noetherian ring R with rt kd R = 1 and lkd R   1?

In either case, lkdR could conceivably be as large as the first contable ordinal!

3. If R is a prime Noetherian PI ring with kd R = 1, P a nonzero prime ideal and   the torsion theory determined by E(R/P) the injective hull of the right R-module R/P is   perfect? Of course such a P need not be localizable.

By Beachy [74], P is localizable if and only if σ is perfect and P_σ, the extension of P to R_σ is an ideal of R_σ. The following could prove to be useful.

Theorem (Gordon): Same hypothesis as in 1. If R is a domain, then σ is perfect if and only if for each essential right ideal I of R such that R/I is -torsion, there exists a central regular element C and an ideal J or F such that c ε IJ and U/cR is σ-torsion.

4. Let R be right Noetherian and P a nonzero prime ideal of R. Suppose each x ε E(R/P) is annihilated by some power of P. Is P localizable?

If P satisfies the AR property this condition is known to be satisfied but not conversely. However, this condition does imply that P is localizable whenever R is an FBN ring or the two-sided ideals of R are principal on the right.

[74] Beachy, J., "Perfect quotient functors," Comm. Algebra 2(1974), 403-427.

Thomas Lenagan:

R will always denote a two-sided Noetherian ring.

1. Suppose kdR = kdI for every right ideal I. Does R have an Artinian classical quotient ring?

Question 1 if of course a generalization of Gordon's result for FBN rings (see Gordon [75]). Question 1 is also related to a result of Jategaonkar [74] which states that if R is an FBN ring and I an ideal of R, lkdR = rt. kdR.

2. (Main Question). If R has a right ideal I with kdI = α, does this imply that R has a left ideal J with kdJ = α?

Ginn and Moss have pointed out another consequence of Theorem (Lenagan [76]): if R has an essential right socle, R is Artin. The obvious generalization of this is the following.

3. If R contains an essential ideal I with kdI = α, does this ordinal fix the Krull dimension of R? Specifically does this imply that kdR = α?

Jategaonkar [74] has shown that if R is FBN, M_R finitely generated with Soc (M_R) essential in M_R, M_R is Artin. Thus, one should ask,

4. If M_R is finitely generated and Soc(M_R) is an essential submodule of M_R, does this imply that M_R is Artinian?

Caveat: A positive answer to 4 implies the validity of the Jacobson conjecture for two-sided Noetherian rings. Of course 4 is false for one-sided Noetherian rings. Thus 4 seems to be a very difficult problem indeed!

5. Can one find a large class of rings for which 4 has an affirmative answer? For example, if G is a finitely generated group and R = Z[G], 4 is true for finitely generated R-modules. More generally, if the ideals of R satisfy the AR property, 4 is satisfied.

[75] Gordon, R., "Artinian quotient rings of FBN rings," J. of Algebra, 35(1975), 304-307.

[74] Jategaonkar, A. V., "Fully bounded Noetherian rings," J. of Algebra, 30(1974), 103-121.

[76] Lenagan, T., "Artinian quotient rings of Macaulay rings," Proceedings, Noncommutative Ring Theory Conference, Kent State University.

Robert Miller:

1. Let $\mathcal{F}$ be a hereditary torsion theory. Characterize those for which homomorphic images of $\mathcal{F}$-injective modules split. Such a $\mathcal{F}$ is necessarily stable.

The above property holds for any commutative domain R with $\mathcal{F}$ the usual torsion theory (Matlis [60]), and for Goldie rings if $\mathcal{F}$ is the Goldie torsion theory (Armendariz [70]).

2. If T is an idempotent ideal of a ring R when is T the trace ideal of a finitely generated projective module?

[70] E. P. Armendariz, "On finite-dimensional torsion-free modules and rings," Proc. Amer. Math. Soc. 24(1970) 566-571.

<u>B. J. Mueller</u>:

Throughout, S will denote a non-trivial semiprime ideal of a (generally) two-sided Noetherian ring R. For relevant definitions and terminology, consult Mueller [74].

1. If S is localizable, does this imply that S is classical? More generally, does there exist a two-sided Noetherian semilocal ring R such that J = J(R), the Jacobson radical of R, does not satisfy the AR property? Note that the AR-property implies that $\cap_{n\geq 1} J^n = 0$. Conjecture, No!

These concepts are readily seen to be equivalent for FBN and HNP rings. See Mueller [74].

2. If $\hat{R}_S$ denotes the $J(R_S)$-adic completion of R_S, is $\hat{R}_S$ necessarily Noetherian? More generally, can one find a Noetherian semilocal ring whose J-adic completion is not Noetherian?

3. Does $\hat{R}_S$ inherit the AR property for $J(\hat{R}_S)$ from R_S?

4. Determine all cycles of any Noetherian ring R. Specifically, determine

<u>Case I</u>. All R having only trivial cycles, i.e., all cycles are of the form {P}, P a prime.

<u>Examples</u>: All commutative rings; separable algebras; classical maximal orders over a Dedekind domain, A[G] with A a commutative Noetherian and G finite and nilpotent; enveloping algebras of finite dimensional nilpotent Lie algebras having characteristic zero.

<u>Case II</u>. All R having enough cycles, but some nontrivial.

<u>Examples</u>: Bounded HNP rings. In this case the cycles are the usual cycles; Algebras finite over their center.

Case III. Which primes belong to a cycle and which do not.

Examples: Enveloping algebras of solvable non-nilpotent Lie algebras. In this case there always exists a prime P which does not belong to any cycle. Other examples which are PI algebras can be found in Mueller [75].

5. Let R be the enveloping algebra of a solvable non-nilpotent Lie algebra. Is it true that each prime ideal P with kdp $< \infty$ ($\equiv$ kdP $= 1$) is nonlocalizable and moreover does not belong to any cycle of R? (See 4., Case III). Conjecture, Yes!

6. Determine all R for which it is true that whenever P belongs to a cycle and Q is a prime ideal with Q P, Q belongs to a cycle.

In general, this property fails to hold. For counter-examples, consult Mueller [75].

7. If R is a prime Noetherian PI ring with kdR = 1, need R be finitely generated over its center?

[74] Mueller, B. J., "Localization of noncommutative Noetherian rings of semiprime ideals," McMaster University Pub., 1974.

[75] __________, "Localization in fully bounded Noetherian rings," ibid., Math. Report No. 78, 1975.

J. C. Robson:

Does there exist a simple Noetherian ring with infinite Krull dimension?

Mark Tepley:

1. For which rings R does Z(M) split? Cateforis-Sandomierski [68] characterized commutative splitting rings. For non-commutative R, Goodearl, in a series of papers (see Teply [76]), reduced the problem to a case of a semiprime ring with zero socle. Subsequently, Teply [76] reduced the problem to a prime ring with zero socle.

2. Find more examples of simple rings.

3. If R is a simple splitting ring, is

a. R left or right hereditary?

b. R left or right Noetherian?

c. If R is right splitting does this imply that R is left splitting?

4. If R is a prime splitting ring, is R left or right Goldie?

There are examples using idealizers (of a semimaximal ideal in a simple ring) which are left and right Ore but not left or right hereditary or Noetherian.

5. If I is an ideal of a prime splitting ring R with Soc $(R_R) = 0$ when is $_R I$ finitely generated and projective? Remark: $_V I$ is already flat.

6. Given a prime splitting ring R with $\mathrm{Soc}(R_R) = 0$, is R an iterated subidealizer of a simple ring S? More generally, does there exist a simple ring S together with a chain of subrings

$$S \supset R_1 \supset R_2 \supset \ldots \supset R_n$$

such that R_i is a subidealizer of some right ideal of R_{i-1} and such that $R_n = R$ for some $n > 0$?

$Z(M_R)$ is said to have <u>bounded</u> <u>order</u> if there exists an essential right ideal I of R such that Z(M) can be embedded in a coproduct of a finite number of copies of $(R/I)_R$.

7. If $Z(M_R)$ has bounded order when does M_R split? More generally, determine all prime rings R for which each M_R having $Z(M_R)$ of bounded order, splits. If R satisfies this property, we say that R has the bounded splitting property or simply BSP.

For sommutative domains R, R has BSP if and only if R is a Dedekind domain. In general, when R has "lots" of two-sided ideals much is known about the structure of R (e.g., see Goodearl [72]). Consequently, one should consider the case where R has relatively "few" ideals or even the simple case.

[68] Cateforis, V. C., and Sandomierski, F. L., "The singular submodule splits off," J. of Algebra 10(1968), 149-165.

[72] Goodearl, K. R., "Singular torsion and the splitting properties," Memoirs of the Amer. Math. Soc., Number 124, Amer. Math. Soc., Providence, 1972.

[76] Teply, M., "Semiprime rings with the singular splitting property," Pac. J. Math., to appear.

[76] __________, "Prime singular-splitting rings with finiteness conditions," Proceedings, Noncommutative Ring Theory Conference, Kent State University.